'Those Dirty Miners'
A History of the Kent Coalfield

by

J. P. Hollingsworth

with

'Kent's Last Days of Colliery Steam'
text and photography by
Tom Heavyside

Sinking No.1 shaft, Tilmanstone 1910.

When no one would sit beside a miner on the bus he said,

*'My face may be dirty but my heart is warm and true.
When you're sitting by your fireside.
Think what a miner did for you.'*

Text © 2010 J. P. Hollingsworth
First Published in the United Kingdom, 2010
by Stenlake Publishing Limited
54-58 Mill Square, Catrine, KA5 6RD
Tel: 01290 551122
www.stenlake.co.uk

Printed by
Blissetts, Unit E1-E8 Shield Drive, West Cross Ind Pk, Brentford, TW8 9EX

ISBN 9781840335088

Contents

	PAGE
AN INDUSTRY IN THE MAKING	3-4

SHAKESPEARE COLLIERY
The Coal Mine on the Beach..................5-12

LOST COLLIERIES..................13-24

Adisham..................13
Cobham..................13
Guilford..................13-18
Woodnesborough..................19
Maydensole..................19-20
Stonehall..................21
Wingham..................22
Goodnestone..................23
Fredville..................24

OTHER SITES OF EXPLORATION..................24

FOUR PRODUCTIVE COLLIERIES..................25-92

Tilmanstone..................25-41
Snowdown..................43-61
Chislet..................70-79
Betteshanger..................80-93

KENT'S LAST DAYS OF COLLIERY STEAM
by TOM HEAVYSIDE
..................62-69

A SHORT BIOGRAPHY OF ARTHUR BURR
..................94-95

A SHORT BIOGRAPHY OF RICHARD TILDEN SMITH
..................96

AN INDUSTRY IN THE MAKING

The history of the Kent coal fields is unique among all the coal producing areas of Britain and as such it needs to be told in a detailed and thorough way. The Kent coal field and its inception are unrelated to the well-established coal fields of northern Britain and South Wales. The only thing it has in common with the more well-known coal fields is the men who worked in them. Coal had been mined in Wales, northern England and Scotland since the beginning of the Industrial Revolution of the 18th century. In fact coal had been used as far back as the Roman occupation of Britain. The coal industry in Kent was a latecomer to the mining trade.

For much of its history Kent has been known as a rural county with a strong emphasis placed on fruit such as apples and cherries and the cultivation of soft fruit such as gooseberries, blackcurrants and strawberries. The growing of hops has also formed an important part of Kent's economic survival while on the Romney Marshes sheep, valued for their flesh and fleeces, have grazed from time immemorial. The Isle of Thanet in east Kent must surely rate as the brassica capital of the country with its large fields of cabbage, broccoli and sprouts. While all these peaceful, rural pursuits were being celebrated above ground there was another 'crop' patiently biding its time beneath the feet of all those generations of farmers. Coal.

Coal began being laid down beneath the fields of Kent during the Palaeozoic era from 542 to around 251 million years ago. Life at the beginning of the Palaeozoic was in the form of algae, sponges and trilobites. In the late Palaeozoic great forests of primitive vegetation thrived throughout what is now North America and Europe. It was then that the great coal seams were laid down. The Palaeozoic was followed by the Mesozoic lasting some 180 million years and subdivided into Triassic, Jurassic and Cretaceous periods. It was during that time that many species became extinct by a mixture of climate change, variations in sea levels and the movement of tectonic plates. The end product of all these dramatic changes resulted in the up thrust of the Wealden dome, a Mesozoic structure lying on a Palaeozoic foundation. Those components created an ideal geological location for the formation of coal beneath the land that would many millions of years later become the area we now know as east Kent. Specifically the area rich in coal seams forms a rough triangle if one is to draw a line joining the towns of Canterbury, Deal and Dover.

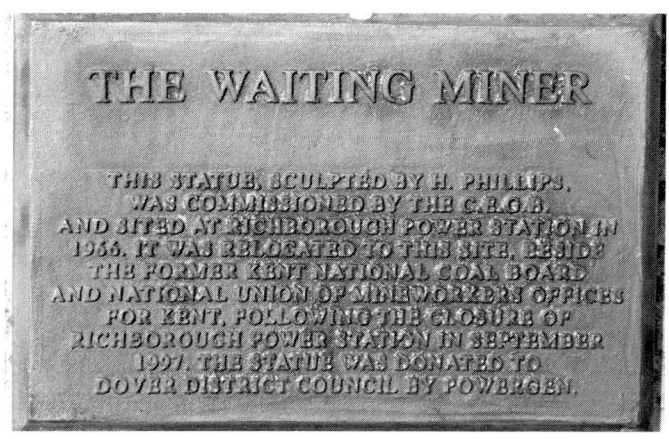

Although coal had been mined commercially in Britain from as early as the 1740s the use of coal as a substance that could be burned to create heat was most likely known by the Romans and Anglo-Saxons. At that time coal was not mined but collected at or near the surface on the beach or along the banks of a river, particularly the Tyne. As early as 1050 AD it had been given the name *saecol* or sea-coal and that name 'stuck' even into the eighteenth century.

When pits were dug to extract coal the name used for it was '*pit-coale*', but because in the south of England coal was brought from the northeast by ships the term 'sea-coal' was adopted not because it came from the sea but was transported by sea. In 1826 a report was compiled by Buckland and Conybeare suggesting that there was a similarity between the geology found in the Department du Nord and Belgium and that found across southern England. Twenty years later Sir Henry de la Beche reached the same conclusion. His work was published under the title, *Memoir of the Geological Survey* in 1846.

In spite of the knowledge gleaned by de la Beche little or nothing of practical value had been done to explore the possibility of coal extraction in Kent. There had been committees formed with lots of talk but no action. On the other side of the English Channel the output of coal mines in the Pas de Calais had already reached 4,000,000 tons a year and the industry employed some 25,000 men. When word spread about the discovery of coal at the site of the channel tunnel workings at the base of Shakespeare Cliff just to the west of Dover it caused great excitement in London where a meeting was at once held. The purpose of the meeting was to work out how best this bonanza so close to the capital could be turned to a financial advantage.

'Waiting Miner' on Dover sea front, 2008.

AN INDUSTRY IN THE MAKING

In 1890, William Morris wrote an article about the news of the discovery of coal in Kent. William Morris (1834-1896) was an artist, writer, furniture and textile designer, poet and founder, in 1894, of the Socialist League. He worked closely with Eleanor Marx (daughter of Karl Marx) and Friedrich Engels. As a socialist Morris saw that such a discovery could lead to the creation of another Black Country among the rural peace and beauty of Kent. He also knew well enough that if coal was mined in sufficient quantities such early speculators as Sir Edward Watkin would care little for the workers who mined it but a great deal about the profits it would generate to him and the landowners. Morris' view was that whatever loss the working man would suffer from the establishment of coal mines would never be compensated for by the 'manufacturing hell' that would be visited upon Kent. He went on to say that: *They, [the workers] have learned by this time that Sir Edward Watkin and his pals will stick to whatever swag they may filch out of Kentish coal, which belongs to the people not to them, and will only yield to the workers what they are compelled to yield.'* He ended by saying that, he hoped, *that coal in Kent will turn out an empty scare*. Fortunately for many the development of the Kent coal field was nowhere near as bad as William Morris predicted. This book tells the history of coal mining in Kent, but that story cannot be told without also telling the story of those who followed the coal to the south east from established coal mining areas in other parts of the British Isles. Behind every miner was a family uprooted from all they knew and deposited like so much luggage in a strange place where the people spoke with different accents, lived different lives and for the most part strongly resented the arrival of these 'foreigners'. And it wasn't just learning to live with 'southerners'. The miners and their families themselves came from very diverse parts of the country and brought their own customs and dialects with them. Families from the Welsh valleys found themselves living side by side with Geordies from the north east and men and women who had spent their lives in established mining communities in Yorkshire and Lancashire. They were all different and proud of those differences. No wonder they resented being lumped together by the locals as 'those dirty miners'.

In the early days, before housing had been built close to the pits, the men had to find lodgings away from their place of work. In the adjacent towns of Deal, Dover and Folkestone landladies did not want to rent rooms to miners. Those miners who were lucky enough to obtain lodgings found it an advantage to keep their occupations a secret and travel to work in ordinary clothes, changing when they reached the pit. In the early days this would also involve washing in some handy nearby stream after their shift because pithead baths were not introduced until the late 1920s. Some who were revealed as miners were given notice and had to leave their lodging houses. A story is told of a miner who had been evicted from his lodgings in Deal. He immediately went to where the fishermen were selling their catch on the beach and bought a fish. This he carried back to the lodging house and without the landlady's knowledge he nailed the fish to the underside of a table where it was left to rot.

In 1915 the first mention is made of the miners organising themselves into a 'trade union'; it was not known by that name but was called the Kent Mineworkers Association. Because there had been no indigenous population of miners the association was made up of men from mining areas across Britain, areas where unions had been in existence for many years. Maybe because many who came to Kent represented the more outspoken and confrontational miners the Kent coal field was always known for its militancy.

In 1919 a book called *The Kent Coalfield* by A E Ritchie gave a detailed account of the Kent Coalfield at the start of the 20th century. The book also prophesised a very rosy future for east Kent. The foreword was published anonymously, but given the address Castle Avenue, Dover was very obviously written by Arthur Burr. His view of east Kent saw it as, *not only collieries, but steel works on a large scale will be in operation in the near future; for the Channel Steel Company has been formed by a powerful group, as will be seen hereafter to work the extensive beds of iron ore ... beds of valuable fire-clay, from which specimen pieces of high grade pottery ware have been made, giving promise of the establishment in due course of manufactories of all classes...*

By 1925 the government had plans to open eighteen pits in the area of east Kent. That would have resulted in three small villages becoming 'new towns' with the total number of houses topping 55,000. The villages earmarked to become new towns were, Wingham, (present population 1,618), which was destined to become a town of 20,000 residents, Woodnesborough, (present population, 400), which would have become a town of 12,000 residents and the tiny hamlet of Ham whose population at the time was less than 100 souls would have grown to a town of 31,000. The plans never reached fruition and Wingham, Woodnesborough and Ham remain attractive, traditional, small Kentish villages. These grandiose plans were made much later than when, and in a different location to where, the story of the Kent coal fields really began. From the first serious exploration for sources of coal, through the years during which the collieries were in production, until the final closure of the Kent coal fields was achieved is little less than 100 years. In that time the coal industry in east Kent moved from privately owned and often poorly equipped pits to a nationalised industry; from groups of disparate people putting down roots in a new area to the present third and fourth generations who are fully absorbed residents of Kent, through various industrial disputes that culminated with the disastrous strike of 1984-85 which led very swiftly to the closure and demolition of the mine workings.

As one chapter comes to an end so another page is turned and the word spreading through the pit villages in the 1990s was 'regeneration'. This is still being pursued with enthusiasm, tempered by a little sadness for what used to be. Although coal mining provided reliable employment for many so the same men are almost all of the opinion that they are happy their sons will not have to go underground at the young age of fourteen and endure the conditions and dangers that they did.

SHAKESPEARE COLLIERY

The Coal Mine on the Beach

Aerial view of Dover Colliery.

After some unsuccessful borings had been put down in Sussex and west Kent, Sir Edward Watkin, on behalf of the Channel Tunnel Company (whose tunnelling under the sea had been vetoed by the Board of Trade), in 1886 undertook to further test the theory of the geologists by a boring on the west side of Shakespeare Cliff, Dover. Mr. Francis Brady, the engineer of the South-Eastern Railway, conducted operations under the geological supervision of Professor Boyd Dawkins. That boring was successful, striking coal measures in February, 1890, at a depth of 1,100 feet, and between that depth and 2,274 feet, where the boring ceased, 14 seams of coal were met with, varying from 6 inches to 4 feet, a total thickness of 23 feet 9 inches, distributed through 1,173 feet of coal measures. This discovery was regarded as of great national importance, for, although some of the upper seams were thin and shaley, lower down the beds seemed richer, the deepest seam being 4 feet thick.

No attempt to make a practical use of this discovery was made until 1896 when the sinking of the Brady Pit was commenced by Mr Francis Brady 280 feet westward of the borehole. Mr. Brady, acting on behalf of the Channel Tunnel Company, carried down that shaft, and then in July, 1896, the Kent Coal Syndicate, promoted by Mr. Arthur Burr, took the matter in hand, the late Mr. George F. Fry, of Dover being the Chairman of the Syndicate, and Mr. Simpson, F.R.G.S., Managing Director.

Water was found in great abundance in the Brady Pit, at a depth of 360 feet, which suspended operations; and, while waiting for the pumps, the second shaft, called the Simpson Pit was commenced in the autumn of 1896, situated midway between the Brady Pit and the borehole. The Brady Pit, owing to the running sand, was eventually lost at a depth of 520 feet. The Simpson Pit (afterwards known as No. 2) was carried down to a depth of 303 feet, when an inrush of water from the greensand suddenly engulfed the men working in the bottom, and eight of them were drowned. This sad accident, which occurred on the 6th of March, 1897, caused delay, but, after the pumps had been put in, sinking was resumed. As a substitute for the lost Brady Pit, another shaft was commenced over the borehole in February, 1898; and to cope with the water, at a depth of 310 feet, a tunnel was driven between the two pits to form a lodgement for pumping purposes. The total water that had to be dealt with at a depth of 450 feet was 54,170 gallons per hour, of which 1,100 gallons was top water, 27,810 gallons from the greensand and Hastings beds, and 25,260 gallons came up the borehole from below 450 feet.

With this amount of water coming in, the sinking was tedious. In the year 1899 progress became very slow, financial difficulties having intervened, but during the sinking a thick bed of ironstone was found, imparting additional value to the coal field. After re-construction of the Company, and changes in the administration, the sinking was continued, but eventually, before coal measures were reached the increasing water stopped the sinking.

An extract from *Dover, a Perambulation of the Town, Port and Fortress*
by John Bavington Jones (published by the *Dover Express* 1907)

SHAKESPEARE COLLIERY

It was the excavations begun to build a channel tunnel that led to the discovery of coal in east Kent. Sir Edward William Watkin had been working with French contractor Alexandre Lavally. They formed The Anglo-French Submarine Railway Company, and their plan had been to tunnel, working from both sides of the channel at the same time. A pilot tunnel was dug from below Shakespeare Cliff, Dover, in 1880-83 using tunnel boring machines that had been patented in 1875, but was abandoned because both the government and public opinion felt that a tunnel joining Britain to the Continent would compromise national security.

THE KENT COALFIELD.

"THE KENT COAL HOLE."
Finding coal in the Channel Tunnel Works. Rush of S.F.R. shareholders to Shakespeare Cliff.
" . . . . half way down
Hangs one that gathers samphire, dreadful trade!"—KING LEAR.

In 1896 the Kent Coalfields Syndicate Ltd. formed by Arthur Burr took over from Watkin. This was to be the first of many enterprises undertaken by Arthur Burr over the next thirty or so years. Shakespeare Colliery was always going to be different; it was not inland but actually located on the beach just west of the harbour below the White Cliffs of Dover, between the towns of Folkestone and Dover. To keep the colliery from flooding a method called 'tubbing' was later employed to line the shaft with steel segments. It was only rarely used in Britain and the name of the method was called 'Kind-Chaudron'. This was a process invented in the 1840s by Karl Gotthelf Kind when he was mining for salt in Thuringia, Saxony. Chaudron is French for cauldron.

View of Dover Colliery from around the 1920s/30s. Large chimney rose above boilers producing steam power for winding gear and boring machinery.

Following Burr's taking over from Watkin an initial shaft was sunk in 1896, but encountered water at 366 feet. The following year the second shaft was begun. It too hit water at 303 feet. Neither shaft was equipped with pumps. These seams were identical geologically to seams located in Somerset, Northern France and Belgium. Consolidated Kent Collieries Syndicate was formed by Burr in 1899, taking over from his earlier Kent Collieries Syndicate. Although now difficult to confirm in the literature, this new company was undoubtedly yet another of Arthur Burr's devious manoeuvres to attract more shareholders and their money to his undertakings; a manoevre not without financial success as there were still many speculators ready to invest their money in what appeared to be an economically viable business. Meanwhile further bore holes were being put down at Pluckley and Brabourne and the Mid Kent Coal Syndicate was boring at Penshurst while the Kent Coal Exploration Company was working at various other sites in Kent.

In March of 1897 a group of fourteen men working the night shift at the Shakespeare Colliery encountered a sudden and unexpected flood of water that rapidly rose in No. 2 shaft to a depth of eighty feet. Eight men were drowned and it was a month before their bodies could be retrieved because of the shaft being flooded and no pumps being available. Of the eight who died only two were natives of Dover, while three of the men came from the traditional mining areas of Nottinghamshire, Derbyshire and Shropshire. The water was thought to have seeped through from fresh water springs as there was no salt found in it and that Nos. 1 and 2 shafts had been dug so close together that the water pressure in abandoned shaft 1 had built up and broken through into No. 2 shaft. Pumps were not installed until it became a necessity in order to remove the bodies of the eight drowned miners. The inquest on the drowned men confirmed the opinion that the shafts had been too close together and the pressure of the water in No. 1 caused the wall to give way. This forced Burr to use steel 'tubbing' on all pit shafts to prevent collapse.

When subsequent bore holes sunk to the west of Dover did not show any coal present the companies found themselves in financial difficulties. The three companies were taken over by the Consolidated Kent Collieries Corporation in 1899. Shakespeare Colliery was never a financial success.

Exploration still continued at other sites in east Kent however and by sinking a bore hole in March of 1900 coal was discovered at Barham, a village south of Canterbury. In June of the same year a noisy and disruptive meeting of The Kent Coal Finance and Development Company was held in London. The outcome was that a committee of five members was appointed to investigate the affairs of the company. Meanwhile, at the August 1900 half yearly meeting of the shareholders of the Channel Tunnel Company it was stated that the Company also held 37,250 shares in the Kent Collieries Corporation, still a viable company with expectations of generating income for its shareholders in the future.

View of Dover Colliery showing miner's houses bottom right.

Dover Colliery

Photo taken from a vessel at sea approaching Dover Colliery from the English Channel *circa* 1900.

The next month a large deposit of iron ore was found at Shakespeare Colliery at a depth of 600 feet. The seam was twelve feet thick and considered to be commercially viable. Unfortunately that commercial viability was never exploited because there was insufficient coal mined to supply the furnaces that would have been required for smelting the ore. In the closing months of 1906 the financial editor of the *Daily Mail* strongly advised his readers to invest in Kent Collieries Ltd. However, in April of 1907 it was admitted that 'circumstances' had retarded the commencement of the output of coal and by September of the same year it was noted that an awful lot more work had to be done before the enterprise could be considered a commercial reality. All this information had been known by Kent Collieries Ltd. of course because by 1905 only twelve tons had been mined, and even by 1907 it was only producing around eight tons a day, less than the mine needed to run the operation and coal had to be purchased from elsewhere to provide power to run the equipment. It took almost twenty years before coal was to be produced in commercial quantities and even then it was never of the best quality. The first commercial load of coal was sold to Leney's Brewery, a local business in Dover, that subsequently advertised that their beer was 'brewed with Kent coal'.

Linings being put into shaft to help prevent flooding.

SHAKESPEARE COLLIERY

Twelve tons raised from Shakespeare Colliery Feb, 1905.

The information regarding the poor quality and quantity of the coal being mined at Shakespeare Colliery did not become common knowledge and people were still being encouraged to invest and buy shares in the company. Consequently the director's report published in May 1908 told the long-suffering shareholders, some of whom had been involved with the company for the past eleven years: *The financial position of the company has recently been a source of anxiety to your directors*. The solution to this much understated sentence was a proposition put forward by the directors whereby 546,000 unissued shares priced at 5/- would be converted into 136,500 preference shares of £1 each. At the same time 163,500 additional shares were issued making 300,000 in all. The owners of the newly issued shares were to be entitled to a 10% preferential cumulative dividend to 80% of the remaining divisible profits. The scheme was approved but did not go smoothly with the directors resigning early on in the debacle and by February 1909 the share price had dropped to threepence each. The Shakespeare Colliery struggled on under other companies but was never a success. In the 1913-1914 *Ward Locke Guide for Dover* Shakespeare Cliff is mentioned as almost a must see tourist attraction: *it ought to be included in the programme of every sojourner in Dover*. The guide goes on to tell the reader that, *'close by the spot are works of the much talked of Kent Collieries. Strangers may approach and view the operations in connection with the project'*. In its short history only a total of 1,000 tons of coal had been mined there. The continuing problems caused by flooding could not be overcome and Shakespeare Colliery closed just before Christmas 1915. The equipment was dismantled and sold for scrap in 1918. The disastrous outcome of the Shakespeare Cliff Colliery was that over an eighteen year period eight men had lost their lives, and a million pounds had been spent with nothing to show for it. The site of the Shakespeare Colliery is still marked on the 1938 OS map of east Kent, it is shown to be sited at the base of the cliffs just where the main line railway emerges from Shakespeare Tunnel west of Shakespeare Cliff.

View showing liners waiting to be inserted into shaft, *circa* 1900s.

With the construction of the present successful channel tunnel in 1987 spoil was deposited beneath Shakespeare Cliff, wild flower and grass seeds were sown and Samphire Hoe was born. It can be reached from the A20 and provides sea angling from over a mile of sea wall, quiet walks where wildlife and plants can be seen, a wetlands area, and refreshment kiosk. Guided walks are given and activities are regularly organised. Kent may have lost a colliery but it has gained a beautiful nature reserve.

LOST COLLIERIES

At the start of the search for coal in Kent many test bores were drilled; in 1896 no fewer than 45 test bores were carried out. Most of them were on private land where farmers and titled estate owners probably hoped to get rich if coal was found. Twelve or more were unsuccessful and never progressed past the exploratory bore holes. A test bore at Brabourne, near Ashford and therefore out of the triangle of later established pits was carried down to 2,004 feet but all that was found were ancient rocks proving the absence of coal-bearing strata. Another bore hole at Ropersole near Barham, just south of Canterbury went down to 2,129 feet and found coal but only in very thin seams. North west of Dover, at Ellinge, the boring went to 1,686 feet, but coal seams were not reached and the borings were never followed up. As well as Shakespeare Colliery there were a total of more than a dozen within the triangle formed by Deal, Dover and Canterbury that actually made it off the drawing board but out of that number only four collieries became fully developed commercial businesses, with others functioning for short periods of time only.

There were no less than 40 companies involved in sinking bore holes. Many were the same companies resurrected under different names after episodes of financial difficulty and many of the companies came under the umbrella of Arthur Burr who was first seen entering the coal exploration field in 1896. Between July 1897 and April 1900 exploratory bore holes were sunk at Pluckley. The bore holes were not successful in finding coal but with an already established brick works in the area it must be assumed that clay was in abundance. The brick works were subsequently purchased by The Contract Company, one of Arthur Burr's businesses and it was at Pluckley that bricks used in the construction of his other coal mining interests were made. It may be pertinent to note here that Pluckley, a small village to the west of Ashford was in the ownership of the Dering family, Sir Henry Dering, 10th Baronet, appears later in the story of Kent coal.

The search for and the possibility of coal being found in east Kent clearly whetted the financial appetite of many speculators. In January 1913 there was a sale of freehold land in and around the villages of Chislet and Herne. Farmers and others in the vicinity could smell a profit and speculators lined up to buy land in the hope that coal would be found beneath it. The winners were without doubt the original landowners and not the gamblers who speculated and in most cases lost out on their investments.

ADISHAM

Adisham is a small village located some six miles south east of the city of Canterbury. In 1918 the site planned for Adisham Colliery was owned by Adisham Colliery Co. Ltd. and their offices were located at Temple Ewell near Dover. Much land in and around Adisham was owned by Sir Henry Chudleigh Oxenden. No action was taken and the proposed colliery never progressed past the planning stage.

COBHAM

Laying outside the east Kent coal field triangle but still an important mine from a historic point of view coal was obtained at this site by opencast methods in the 18th century. The coal deposits were on the estate of the Earls of Darnley, Cobham Hall. The coal thus obtained, small amounts of brown lignite, had been used by the family over a long period of time to heat the hall. The coal was not taken from mines but dug out close to the surface by workers on the estate and it had never been a commercial enterprise.

In 1947 a shaft was sunk and the mine reputedly produced eighty tons a week. In 1953 the mine was closed and the site completely cleared. The site of the pit is no longer in evidence as most of it lies beneath the A2 between Rochester and Darenth. Cobham Hall still exists and has seen life as a hospital in the First World War, accommodation for fighter pilots in the Second World War and in 1962 it became a girl's school.

GUILFORD COLLIERY

Coldred is a small hamlet to the east of the present A2 that runs between Dover and Canterbury. It carries the proud title of Kent's Best Kept Village and lies roughly between the village of Sheperdswell and Waldershare Park, which was home to the Earl of Guilford. As in the case of some other collieries in the area it was known by a variety of names, Coldred, Guilford and Waldershare, all of which were misnomers. It was not located in the village of Coldred, there is no such place as Guilford and Waldershare was the name of the Earl of Guilford's estate. To further complicate things the colliery was located in Singledge Lane and some people called it Singledge Colliery. Good coal seams had already been located in the parkland on the Earl's estate and he was keen that the exploration should be widened, although still on land that he owned.

Guilford Colliery in advanced stage of construction.

First coal from Guilford number 3 shaft 1905.

The colliery was started by Arthur Burr, a speculator and local business man in 1906. In April of that year Burr registered what was known as the Foncage Syndicate, a subsidiary company of his Kent Coal Concessions Ltd. Because the site was in an isolated, rural location all heavy equipment had to be moved in over fields and unpaved lanes full of ruts and mud. The first manager at the pit was Mark Ramsdale and the undertaking was owned by Guilford and Waldershare Colliery Co. Ltd. with Arthur Burr as managing director. The first shaft was sunk at Guilford in July 1906 and reached a depth of 300 feet. Thick seams were identified and two shafts were sunk each with an eighteen foot diameter. A small shaft at 298 feet was not worked but left to collect water draining from the chalk with the water utilised to work the winding engines.

A total of five seams were identified giving a total of 16 feet of coal plus a large deposit of iron ore and several good beds of fireclay, a valuable commodity at the time. Because of the lack of roads to the site all work stopped during the winter. In spite of an injection of £20,000 there was never enough money to develop the mine to its full potential. The length of time spent before a railway connection was made also did not help its commercial viability.

At the same time that Arthur Burr and his companies were working at Guilford Colliery, shafts were being sunk at nearby Tilmanstone, also under his control and operating under the name Foncage Syndicate. Coal was expected to be mined at Tilmanstone within a few months of sinking the shafts. It was thought that if Tilmanstone began to produce coal in large quantities that would make the purchase of the Guilford operation more tempting to a prospective buyer. However things did not work out that way, there were long delays encountered at Tilmanstone that made the continuation of work at Guilford seem like a good idea. In December 1909 preparations went ahead for the sinking of shaft No. 2 at Guilford. £22,000 of the syndicate's capital had been issued but the expenditure amounted to £40,000. The shortfall was taken up by a loan from the east Kent Contract and Financial Company. Whether this loan ever actually existed may never be known but the director of the company is shown in the 1911 edition of the Blue Book for Dover as none other than Arthur Burr. Shareholders were made aware of the need for a fresh injection of money and were promised large returns on their investments. Arthur Burr was still in overall charge at Guildford and was also sinking shafts and seeing to the construction of surface buildings at other sites throughout east Kent. Because the search for coal in Kent had been going on for more than twenty years without any real results Burr found it difficult to obtain additional funding. In 1910 he formed the Guilford (Waldershare) Coal Fireclay Company Limited, and although it was about to be registered with a capital of £460,000, this did not happen, although the reasons are not known. Other, smaller loans were negotiated and the work at Guilford Colliery continued in a stop start fashion throughout the rest of 1910. Finally by November of 1911 both shafts had been sunk and lined to a depth of 900 feet.

Guilford Colliery

Eventually, in 1912 the East Kent Light Railway connected the site to their station at Eythorne. In 1914 shareholders were made aware that the colliery was encountering severe financial problems. The decision was made to sell the mine as a going concern as it was losing money and a French company paid £150,000 for it. Operations were suspended during the First World War and recommenced in 1918. Throughout the war the shafts had become flooded, but at its end they were pumped out and a new power station built preparatory to recommencing the search for coal. The owners tried to overcome the problem of water entering the shafts by using the cementation process utilised at other sites in the area. By 1919 the search for coal had proved unsuccessful, the cementation process was expensive and did not completely keep out the water. The colliery was finally closed in 1920 and the French pulled out.

In 1929 the colliery was purchased by Richard Tilden Smith, the then owner of Tilmanstone Colliery, and he had some of the equipment moved from Guilford to Tilmanstone. The coal seam found under Waldershare Park was eventually mined from the colliery at Tilmanstone.

East Kent Light Railway station at Eythorne.

Aerial view of Guilford *circa* 1930.

Today there are two buildings left at the site of Guilford Colliery; they are just off the road in Singledge Lane and can be easily seen. The large, red brick winding house has survived but has been converted to a private house. It looks very handsome and well-preserved. Next to it is another brick building from the colliery. The site of the colliery is still marked as such on the 1938 OS map of east Kent.

WOODNESBOROUGH

The site was located just south of the village of Woodnesborough which is situated south of the ancient town of Sandwich. The owners were ostensibly, Intermediate Equipments Ltd. based at Castle Hill House, Dover in 1910. Work began that same year to extract coal from the Kent coal field, but the Woodnesborough site was never a success. The company showed a capital of £100,000. This was yet another company started by Arthur Burr; it had no board of directors. By 1910 there were eight companies working under the umbrella name of Kent Coal Concessions. Arthur Burr, whose name crops up regularly in the coal industry of east Kent was appointed (appointed himself probably) the managing director of the company, and also stood to gain 20% of any profits. The name used for the development at Woodnesborough was the Goodnestone and Woodnesborough Colliery Company Ltd. The site was sometimes referred to as Hammill Pit. Burr boasted that by 1920 there would be at least twenty collieries working in Kent and that each of them would be producing a minimum of 500,000 tons of coal a year and employing 20,000 men.

By 1911 much in the way of building had been completed at the site including workshops and an engine house. No shafts had been sunk or attempted at this time. Any more work was held in abeyance until a branch of the East Kent Light Railway could serve the site. Nothing more happened before 1914 by which time the First World War intervened. Shortly after the war began the buildings were commandeered by the army as a cavalry re-mount unit and were utilised as stables.

In 1919 Arthur Burr died. At the time of his death he was deeply embroiled in many financial problems and it is very likely that had he lived his career as a mining speculator would have been over anyway. In 1920 plans were once again drawn up to construct a colliery at Woodnesborough. These plans were more ambitious than the ones envisaged by Burr and encompassed not only a colliery but housing to accommodate 12,000 people. These plans were abandoned as the start of the Depression put paid to all such expansionist schemes.

Intermediate Equipments Ltd. began to sell off or otherwise dispose of surface plant that had been installed and after the war nothing more was done at the site. In 1923 the would be mine was sold to Pearson & Dorman Long, the owners of Betteshanger Colliery. They held onto the mineral rights but sold the site to the Hammill Brick Co. Some of the buildings already on site were utilised by them and they opened their brick works in 1927. The East Kent Light Railway did open a line in October 1916, but with no activity at the site of the colliery it was never used. When Hammill's bought the site they used the spur to transport some of their bricks though most went by road. The line survived until 1951.

Later another Woodnesborough Colliery was proposed close to the village of Eastry and although land was acquired nothing came of these plans. The National Coal Board sold the land in 1951 and it reverted to agricultural use.

MAYDENSOLE

This colliery was to have been located at East Langdon, a small farming community situated about half a mile north of the Deal to Dover Road, the A258. An advantage to siting any industrial undertaking in that location would have been the close proximity of the railway, the line joining Dover to Deal having been laid in 1880. The intended pit was begun by Burr's Intermediate Equipments Ltd. in 1910. Several bore holes were drilled but work was then stopped without any further exploration being undertaken.

In 1914, in the same vicinity, Arthur Burr with his then partner Sir Henry Edward Dering under the name of Deal and Walmer Coalfield Ltd. leased a seven acre field from a farmer, George Jeken. The lease stated that Deal and Walmer Coalfield Ltd. was to have the mineral rights to the land and would pay George Jeken royalties on any and all coal mined there. A snag occurred when it came to light that Jeken had taken out a mortgage on the land in the amount of £1,250 from Dr. John Pycock Candler of Loughton, Essex. The local water company also got involved because they drew water from an underground source on the site and were concerned that the supply could become contaminated by any industrial activity in the area. With those problems apparently resolved Burr forged ahead and began the construction of various buildings on the site, but in February 1918 there was a supplement to the original lease. This supplement asked that George Jeken should release the lessees from payment of all rents due and accrued from July 1914 up until September 1917 and from all other claims and demands. It was also stated that all monies owed were to be paid to George Jeken three months after the end of the war.

The most interesting change to the new lease was the insertion of just a couple of words. The original lease included the words, 'in the event of Arthur Burr or Sir Henry Edward Dering becoming bankrupt'. However, inserted by hand alongside those words on the new lease and initialled by both Burr and Dering can be seen the words, 'or being'. From that it can only be assumed that once again Burr was short of money and that he had now been declared or was about to be declared bankrupt.

When Richard Tilden Smith purchased Guilford and Tilmanstone Collieries in 1926 he must have also taken up the lease of the land at East Langdon because although he does not appear to have planned to mine the land there he certainly made use of it later during his ownership of Tilmanstone Colliery.

STONEHALL

The end of Stonehall.

This very short-lived colliery was located near Lydden a small village lying to the west of what is now the A2 Dover to Canterbury road. It was begun in 1913 and was owned by Stonehill Colliery Ltd., Temple Ewell, Dover. The people behind the name were two Frenchmen. At the start of the First World War they more or less abandoned their enterprise and returned to France. The original manager was A J Kennedy. A total of three shafts were dug there, the north to a depth of only 75 feet and the east and west shafts to a depth of 273 feet each. The chimney at Stonehall stood at a height of 180 feet, 30 feet higher than the one at Snowdown Colliery. In 1914 all work was stopped as no coal had been found. In 1919 work restarted but by 1921 all work ceased again and most of the buildings and equipment were demolished.

Once again there was, and is, a muddle over names as some people knew this colliery as Lydden Colliery; it is on the very edge of the village of that name, while some called it Stonehill and some Stonehall. The remaining buildings may be seen from the end of a cul-de-sac in the village called Stonehall Close. A long, redbrick building is all that remains of the colliery. It sits in a large field in an idyllically beautiful valley that rises up to the height of 400 feet on its east side. The building looks well-preserved and appears to be used for agricultural purposes. On the west side of the valley the rail line still runs between Dover and Canterbury.

View of Stonehall with workers houses in the mid-distance.

The drilling rig on the right of the photo was used during the 1920s and 1930s to investigate the possibility of freezing the strata to prevent flooding in the shafts.

Last building left at Stonehall Colliery, 2008.

WINGHAM

Wingham is a busy, prosperous and historic village located on the A257 between Sandwich and Canterbury. The construction of a colliery at Wingham was begun in 1910 by a company known as the Wingham and Stour Valley Colliery Ltd. and was yet another company headed by Arthur Burr. Two shafts were sunk but as with all collieries in the east Kent coalfield there were problems with flooding and no money available to purchase pumps. All work was stopped in 1914. The buildings above ground were 'mothballed' and finally sold in 1924 to Grain Harvesters, a milling company for agricultural feed stuffs. Railway lines had been laid in anticipation of the coal that never materialised with three stations, Wingham Colliery, Wingham Town, and Wingham Canterbury Road all built by the East Kent Light Railway.

Eastry Junction

Canterbury Road Station

The line was a failure and closed to passengers in 1948 followed in 1951 by the closure of a section that ran north to the village of Eythorne. Further rail plans had been drawn up in the 1930s and preliminary work was started on a line to run from the Wingham Canterbury Road station to Chislet Colliery via Stodmarsh, but these plans did not come to fruition. Today the site of Wingham Colliery Station is still in the ownership of Grain Harvesters.

Goodnestone

This small village is located some two miles south of the A257 Canterbury to Sandwich road. Goodnestone Park dates from 1704. The owner was related by marriage to Jane Austen who often visited the estate. The exploration for coal was carried out within the park boundaries. In December 1906 a third boring was carried out at this site in the eastern corner of Goodnestone Park. The bore went down to almost 1,000 feet with a diameter of eighteen inches. That made it possible for the surveyors to carry their exploration down to the lower coal measures which were as deep as 4,000 feet. The chalk at Goodnestone was much thinner than had been found at the site of the borings that had been made at Waldershare, site of the Guilford borings and the geologists at the time were very optimistic of finding good and commercially viable coal seams. Alas, their predictions did not come to fruition and once more as on other sites of early exploration the hope of prosperity for Brook Bridges, the owner of Goodnestone Park did not materialise. Although not being developed for coal production the house and grounds remain a popular destination for those wishing to tour stately homes.

Goodnestone Park

FREDVILLE

This was the site of an early bore hole; the land was on the estate of Mr. H.W. Plumptre who lived at Fredville Park. Located close to (the eventually successful) Snowdown Colliery, between the villages of Shepherdswell and Adisham, the second boring at Fredville reached coal seams at 1,363 feet. The three seams gave a total of seven feet four inches. As at Guilford, (Waldershare) there were also deposits of iron ore and fire clay. Coal was reached at an average depth of 1,458 feet. Exploration at Fredville continued into 1908 but was not further developed. It is probable that lack of finance was once again the reason.

Fredville, 1906, site of the future Tilmanstone Colliery. Arthur Burr, extreme left with Malcolm Burr his son next to him.

OTHER SITES OF COAL EXPLORATION IN EAST KENT

In addition to collieries that almost got off the ground 'lost' test bores were drilled at other sites. Coal was found at four of the sites but no further work was done. These sites were at Barham, just south of Canterbury where a bore was sunk to 2,129 feet and located twelve thin seams. At Waldershare, some four miles north of Dover, the bore went to 2,372 feet and five seams were found, and at Nonnington a bore of 1,505 feet revealed three seams. At Goodnestone, south east of Wingham, no coal was found following a test bore to a depth of nearly 1,000 feet but it was believed there were seams at a depth of 4,000 feet. The Anglo-Westphalian Company drilled bore holes at Hoades Wood, Sturry; Reculver, Herne Bay; Chitty and Chislet Park while exploring for coal before Chislet Colliery was started. The problems encountered not only by the owners of these failed pits but by the ones that subsequently went into long-term production were how to get the coal from the coal face to the customer and where to accommodate the workers and their families.

Without exception the pits were located in isolated rural areas where roads were bad or non-existent and the small nearby villages and farms could not and would not cope with the influx of people, many of whom came from traditional mining areas in Great Britain and were seen with suspicion and even fear by the local population. In a time when population movement was much more restricted than it is now these new arrivals were seen as 'foreigners' by the local Kentish people. East Kent as a predominantly rural area had not been exposed to the influx of workers from other parts of the country as had London or even the large towns like Rochester or Chatham. Indeed as with small villages everywhere there was not the infrastructure needed to support a large influx of people. Schools, shops, churches, medical care, none of these vital services existed in large enough quantities, if at all in the ancient villages of east Kent.

FOUR PRODUCTIVE PITS

When the 'Kent Coal Rush' finally slowed down and the speculators went away to lick their wounds the Kent coalfield was born and grew into an industry that would flourish and bring financial benefits to what had been a quiet and pastoral part of Kent. The big difference between the Kentish coalfield and the northern pits was that somehow Kent never lost its rural aspect. Coal mines existed next to fields of wheat and fruit orchards. Kent never became an area of those 'dark satanic mills' of the heavily industrialised parts of Britain such as South Wales or Yorkshire. The miners for the most part lived not in old poorly-constructed houses but in housing newly built to accommodate the influx of workers and their families. Four collieries, the men who worked them and the families they raised around them eventually became a part of the landscape of east Kent. Through good times and bad everyone thought they would go on producing coal forever until it all came to a sad and bitter end when the axe finally fell in the late 1980s. Men who had known no other occupation lost their source of income, but they had lost more than money. The closeness of a mining community, the mutual support it offered and the social activities the workers and their families took part in were all gone. There was anger and despair; hundreds of men were suddenly out of work in an area that already endured a high unemployment rate compared to the rest of the country. The phrase 'you can't teach an old dog new tricks' was the truth as many of the ex-miners looked at their situation. But all that trouble was a long way in the future when the four pits opened in the first quarter of the twentieth century.

The first pit to go into operation in 1906 was Tilmanstone, some five miles inland from the coastal town of Deal. Snowdown followed in 1908, and Chislet, near Canterbury started production in 1914. The last pit to open in 1924 was Betteshanger located about a mile north east of Tilmanstone.

TILMANSTONE COLLIERY

'The most advanced colliery is at Tilmanstone. This was commenced on the 7th of July, 1906 by the Foncage Syndicate, an offshoot of the Concessions, under the management of Mr. Nathanile Griffith, M.Inst.M.E. It was afterwards transferred, under the same able manager, to a new East Kent Colliery Company. This pit is equipped with engine power for pumping and winding, and is expected to be the first shaft to reach coal'.

John Bavington Jones 1907

No. 2 shaft named for Arthur Burr's grand-daughter Gabrielle (seen in photo), 1907.

Tilmanstone, also known as the East Kent Colliery, was the oldest of the four pits in the Kent coalfield. It dated from 1906 and was located close to the ancient village of Eythorne. The first sod of what would be the No. 2 shaft was turned in November 1907 by one of Arthur Burr's grand-daughters, Gabrielle. The shaft would be named after her. The turning of the first sod was accompanied by much celebration, but unfortunately there was not enough money for the construction and even the winding engine did not arrive until nine months later.

At that time miners were represented by the Miners' Federation of Great Britain, an organisation that had been in existence for just seventeen years. Tilmanstone Colliery was initially owned by Arthur Burr's Foncage Syndicate. It was later managed by another of Burr's companies the East Kent Colliery Co. Ltd. Like other enterprises started by Burr, Tilmanstone became a victim of lack of investment and the ongoing saga of too much underground water, which was either not acknowledged by surveyors or ignored by management. The same mistakes of too much water which had been made at Shakespeare Colliery were repeated at Tilmanstone.

The colliery's first fatal disaster occurred in 1909 when a hoist bucket or hoppit as they were known broke lose and fell down a shaft. Three men were killed and the water pipes from the pumps were fractured. Work had to be abandoned for almost a year after water poured into the pit at a rate of 1,000 gallons per minute. Work on sinking the first two shafts recommenced in 1910 but it was very slow as not only did water continue to cause problems but tons of sand also entered the shaft. No. 3 shaft, the Rowena, named for Burr's second grand-daughter was begun in August of 1910. Finally in 1912 electric pumps were installed.

In 1912 the shaft reached the level of a coal seam and the following year at a depth of 1,560 feet the seam to be known as the Beresford was hit and commercial mining could begin in earnest. When the first coal was raised in November 1912 Burr was apparently very ill but would not forgo seeing the momentous event and had himself carried to the pit head by two burly miners. The owners also wanted to exploit the richer Milyard seam, at a depth of 3,000 feet but lack of finance stopped further development. Also in the same year the East Kent Light Railway from Shepherdswell to Eythorne was improved to permit equipment to be more easily transported to the site.

In 1914 water again caused problems when it broke through and flooded the pit causing the stoppage of all work for two months. When work recommenced the miners had to switch from longwall working to pillar and stall. This was a slower method and as a result output was drastically reduced. The *Dover Express*, full of the potential offered by collieries in the area even ran a column in the paper each week giving its readers regular updates of what had been achieved at sites throughout the area.

It was not only speculators and the stock exchange that saw the potential for profit from the coal mine. The following advertisement appeared regularly in local papers:

Kent Coal Developments
Shopkeepers and others requiring business premises or sites at Eythorne, Wingham or Shepherdswell should secure same without delay. Full requests to: Messrs Flashman & Co. Estate Agents Dover.

Advertisements such as this revealed the great faith and interest that had been engendered by talk of the great wealth that was expected to come from the discovery and mining of coal in east Kent. With the expectations of tons of coal being dug pound signs were dancing in front of the eyes of many local businessmen. No one wanted to be left out of this promised bonanza. In these early days an account of the South Eastern Coalfield, as it was sometimes called appeared in a locally published guide book for east Kent. The article stressed that The Kent Coal Concessions Ltd. and its allied companies had no connection with the 'old' colliery workings at Shakespeare Cliff and went on to inform the reader that samples of coal could be viewed at the Museum, 12a King Street, Dover and boasted of the most modern machinery being used during the course of construction at Tilmanstone, Snowdown, Goodnestone, Woodnesborough and Stonehall Collieries: *Coal in the amounts of 1,000 tons a week has been raised at Snowdown* the article trumpeted and is accompanied by a photo of the workings at Snowdown, Tilmanstone and of Stonehall Village. The article finished by telling the reader that permission to view the sites could be obtained at the offices located at Castle Hill, Dover.

TILMANSTONE COLLIERY

The coal from the Beresford seam was found to be more suitable for industrial than domestic use but this did not stop the company reporting to the media that the Beresford seam at a thickness of four foot eight inches was 'a highly bituminous coal of exceptional quality'. However, the company found that markets for the coal were not easy to find. Estimates of the amount of coal that could be extracted from the mine had been set at 6,000 tons a day but even under maximum production no more than 750 tons a day was obtained. Arthur Burr, always the optimist, announced expected profits of £120,000 for 1914, but by June of that year the receivers moved in. In spite of its early successes the company was in financial trouble and Arthur Burr was fired. Following the departure of Burr the shareholders put together a rescue package and took over day to day management of the colliery. In spite of their efforts the pit still continued to lose money year by year.

Going back to the pre-war years Arthur Burr had realised from the outset the problem of accommodating his workers. Because the workers at Shakespeare Colliery had lived in Dover it had been expected that miners working at Guilford, Tilmanstone and Snowdown would do likewise. Burr however had different thoughts; he wanted his men close to their work. There were little or no rental properties close to the colliery and anyway the locals were not too enthusiastic about having an influx of 'foreign' workers. To help solve the problem of housing 'unattached' miners Burr leased the then unoccupied Elvington Court and set up dormitories for his men. Burr also began to build small estates near the villages of Elvington, Woolage, Snowdown and Stonehall. Elvington Court's first residents came from as far afield as Wales, Yorkshire, Lancashire, Nottinghamshire and Somerset.

Elvington Court (now demolished).

Navvies preparing the ground to lay railway track to Tilmanstone Colliery, *circa* 1915.

Eyethorne Station with Tilmanstone Colliery in the background, *circa* 1960s.

In 1916 the East Kent Light Railway opened a station and named it Tilmanstone Colliery, its aim being to serve the men living at Elvington Court, to carry them to and from work. In 1925 it was renamed Tilmanstone and again in 1927 its name changed to Elvington. There wasn't any road access to the station, just a footpath running from Pike Lane through the colliery to the miners' houses at Elvington. The line was finally closed to passenger traffic in 1948 and closed completely in March of 1951.

Throughout the First World War owners' demands for higher productivity and the repeal of the eight hours legislation was put into practice thereby worsening working conditions. In 1917 the output from the pit was 'reasonable' and for the first time the undertaking showed a small profit. At the beginning of 1918 the chimney was encircled by a large iron band about three-quarters of the way up, and this band was then fixed to the ground by four stays thus giving the structure added strength. In 1919 the MFGB called for a six hour day and a 30% pay increase. The Sankey Commission's findings recommended the pay increase and actually asked for a six hour day for each of the three shifts. It also made the recommendation that the coal industry be nationalised.

The Prime Minster Lloyd George, a Welsh man, who it might be thought would have some sympathy for the miners of his homeland, did not honour the recommendations. That decision however may not have been entirely down to him. At the time he was hanging on to power by the skin of his teeth and was dependent upon the Conservatives for his survival.

Tilmanstone Colliery

Tilmanstone struggled to succeed as a commercial business and it continued to lose money. In 1925 Richard Tilden Smith moved in as manager. He had been interested in obtaining the ownership of Tilmanstone since 1914. In 1926, the year of the General Strike, the colliery again went into receivership. It was at this point that Tilden Smith took over as owner under the company The Tilmanstone (Kent Colliery) Ltd. The General Strike caused high unemployment throughout the British Isles and it was following the strike that miners who had been black-listed for their strike activities found it was no longer possible to work in their home pits. Many of these men travelled from the northern mining areas to Kent to seek work.

Tilden Smith had adverts put in newspapers inviting miners to come to work in Kent. The response was so large that other adverts had to be inserted telling miners that all the vacancies had been filled, but still they came desperate for work at any cost, many walking hundreds of miles, not having the money for train fares. Because of their previous employment problems many miners signed on for employment in Kent using false names. The incoming miners were not well-received in towns such as Deal. The local people saw them as 'dirty' and noisy and they were unable to understand the broad 'Geordie' and other accents that they spoke. Signs stating 'No Miners no Dogs' appeared in lodging house windows and shops.

Unlike Arthur Burr, Tilden Smith planned that Tilmanstone should be a stable, profitable and reliable business undertaking. He was not out to make a 'quick buck' as Burr had been. Unlike Burr, Tilden Smith did not have partners; he had the money to back his plans without having to resort to Burr's somewhat dubious financial methods. Another commercially viable seam the Milyard was hit in May 1930 at 3,035 feet.

Richard Tilden Smith addresses miners during the labour crisis in the 1920s. This resulted in him sending a party of miners to the USSR to study the working and living conditions of miners there.

Following the rise of Communism in Russia there had been talk among the miners, some extolling the near to ideal conditions that Russian miners worked in. Tilden Smith saw that such talk would lead to industrial unrest. His solution was simple and he took immediate action. Two men were selected by the miners, one an avowed communist the other a member of the local Labour party. Along with a local school master the three men travelled to the heart of the Russian coalfield. All expenses for the trip the men took were paid for by Tilden Smith. They went as ordinary travellers with no one in Russia knowing that they were there on what would now be called a fact-finding mission. In the Donetz Basin, the chief mining area of Russia, they were able to study conditions in which the Russians worked and lived. The conclusion reached by the two miners was that if British miners had to endure such conditions, 'there would be hell to pay'. At a meeting held to enable the two men to recount their experiences a notice had been displayed stating that free passage for the miners and their families would be advanced to any miner present who wished to go to work in Russia. No one took up the offer.

Tilden Smith was also ahead of his time in taking an interest in both the accommodation provided for the miners and their families and the facilities available for recreation. The housing he arranged through the Eastry District Council, and a hundred homes were built of varying designs, these houses are still occupied in the village of Elvington. For recreational facilities Tilden Smith made a large barn on his property of Elvington Court available to the miners. While the miners did a lot of the work of converting the barn the workmen that were needed gave of their time. The finished facility had a seating capacity of 750, electric light and heating, a stage with dressing rooms and a dance floor. Along with weekly dances the miners and their families enjoyed boxing tournaments, concerts and an amateur dramatic society was formed. In addition to the barn conversion Tilden Smith provided a building that was used as an ambulance station. He paid for all the equipment needed to staff the station including having his Rolls Royce car converted into an ambulance. This he presented to what would become the first St. John's Ambulance brigade in the Kent coalfield. The ambulance was in regular use until 1956. Later a levy of a penny was taken from each miners' weekly wage to subsidise the brigade. It was increased to 3p in 1971 and continued until the pit closed.

Even in the 1920s accommodation for the miners was essential. Before being joined by their families the miners like many others before them found it difficult to obtain lodgings. To solve the problem Tilden Smith like Burr some fourteen years previously utilised the house, Elvington Court which he now owned and converted it into a hostel for the miners. They were fed and their snap was provided for them to take to work each day.

Children's free meal at Eythorne School during the 1926 General Strike.

Pit workers at Elvington Court, *circa* **1906/7. It was used by the miners as a hotel and social centre before housing was built for them.**

TILMANSTONE COLLIERY

Tilden Smith planned to export coal from the port of Dover but the only method of transportation available at that time was via rail. This method was not only slow but very expensive. To transport one ton of coal just ten miles to the port of Dover cost the company 5/9d. His solution to the problem was to build an aerial steel ropeway by which the coal could move directly from the colliery to Dover. Tilden Smith had long nursed a plan for this type of transportation. His plan was initially turned down by Dover Council but Tilden Smith was given permission by the Dover Harbour Board to have a bunker located on the eastern arm of the docks that could hold 5,000 tons of coal. The Harbour Board contributed £25,000 to the cost and the aerial ropeway was put under construction. When up and running the ropeway, supported by 177 trestles, crossed the countryside from Tilmanstone to Dover. The plans for the ropeway made mention of the fact that the wire cables carrying the coal buckets had to be of a height that would enable the highest hay cart to pass safely beneath it. On its route the aerial ropeway crossed fifteen roads and two railways on its journey to the eastern arm of Dover's docks. The final part of the journey carried the buckets through a quarter mile long twin tunnel in the cliffs before the coal was tipped into a bunker. The buckets of coal containing 14.5 cwt. left Tilmanstone every 21 seconds with the buckets spaced 46 yards apart.

On October 12th 1929 the first section of the aerial ropeway running from the colliery to East Langdon was completed and celebrations featuring Tilden Smith and various officers of his company were recorded on film by the local press. At the celebrations Tilden Smith presented Mrs R J Barwick with a model of the aerial ropeway and she inaugurated the machinery by making a phone call from the site at East Langdon to Tilmanstone requesting that the first bucket of coal should leave from the colliery. In the field was a building that had been converted for use as the power station for the ropeway. A strange coincidence was that the site of the power station was in the same East Langdon field that Arthur Burr had leased for coal exploration in 1910 and 1914.

Tilmanstone Colliery

TILMANSTONE COLLIERY

In February 1930 the first ship, the 'Corminster' arrived to take on a consignment of coal at Dover. The aerial ropeway allowed 500 tons an hour to be loaded. Unfortunately Tilden Smith died before the ropeway could be put into full operation and in spite of its apparent advantages the export of coal through Dover did not meet its expectations. On March 14th 1930, the *SS Cordale* took on 2,500 tons of coal for delivery to the paper mills at Sittingbourne. By coincidence the method of unloading employed at Sittingbourne Docks was also by an aerial ropeway that carried the coal directly to the mills.

Spokesmen for the railways stated that the ropeway was a waste of time and money. They had reduced their rates to 2/- a ton. But would they have still made that reduction without the construction of the ropeway? It was not used much after 1935 and during the war it fell into disrepair and was finally dismantled and sold for scrap in 1954. Following the death of Tilden Smith the business was continued by the board of directors until in 1937 it was sold to the Anglo-French Consolidated Investment Corporation Ltd. who planned to further develop the lower seam to allow the colliery to work at full capacity. In spite of the Second World War, heavy investment was continued and eventually losses were turned into a small profit with a 5% dividend being paid to the shareholders.

View of eastern arm.

Loading machinery for coal on eastern arm.

TILMANSTONE COLLIERY

The construction of Elvington village was continued under Tilden Smith and the Tilmanstone Miners Dwellings Syndicate was formed and built 230 houses. The typical dimensions and facilities to be found in one of the houses were as follows:

Ground Floor	
Living room	13 ft x 11 ft
Parlour	12 ft x 11 ft
Scullery	11 ft x 7 ft
(cooking range, copper and sink)	
Yard, Larder and coal shed also on ground floor.	

First Floor	
Bedroom	12 ft x 11 ft
2nd Bedroom	11 ft x 9 ft
3rd Bedroom	11 ft x 10 ft 6 in.
Bathroom	7 ft x 5 ft 5 in.
(with hot and cold water supply)	

Elvington village

It must be born in mind that these houses were all built pre-Second World War. In the 1951 census for Great Britain the returns showed that across the Durham Coalfield there were still 40% of homes with three rooms or less, 42% without a bath and 26% of homes still had to share lavatory facilities. When it came to washing facilities the figures were about the same right across Yorkshire and Lancashire and the South Wales valleys. The village was eventually well-served by shops. By the early 1930s there was a Co-op butcher, greengrocer and grocery shop plus a garage and a cobbler. A travelling fishmonger also visited regularly. A building known as the 'settlement' was initially erected where leisure, educational and welfare facilities were available. Later there were twice weekly visits from a doctor and a health visitor and a library with more than 600 books. In 1931-1932 the settlement was rebuilt to include kitchen facilities. A male voice choir was formed and athletics, bowling, boxing, cricket, football and tennis teams were very popular, with regular matches played against other teams. In 1959 a new welfare hall was opened and celebrated with a dance; Elvington even had its own dance band. Next door to the welfare hall was the working men's' club. A school, primarily for the children of Tilmanstone miners living in Elvington, was constructed in the 1920s. It was built on the edge of Eythorne Village and is now known as the Eythorne Elvington Community Primary School.

Back at Tilmanstone, one of the deepest pits in the country at approximately 3,000 feet with the seams overlaid with around 900 feet of chalk, the old problems of water seeping in was still an ongoing problem and even with modern pumps it was always known as a very wet pit. It was not unusual for miners to work up to their knees in water in spite of pumps running continually. One particular part of the pit was known sarcastically as the 'high and dry' because it was so wet and low lying. This name has been kept alive by a pub in Eythorne called 'The High and Dry'.

In the early days each miner was given discs or mottes, the number on it identifying a specific miner. These discs would accompany the coal to the surface where the tubs were inspected. If they contained too much 'rubbish' the miner would have his pay docked. Because this could cause disputes between the checker and the miner later on there had to be two checkers one of whom had to have been appointed by the union. The method for extracting the coal was known as pillar and stall*. Two men, a collier and a trammer, worked together. The collier, working with a hand pick would hack out the coal and the trammer would then load it into tubs which when full he would push on rails to a central collection point where as many as 30 tubs would be linked together and moved on by haulage lads. Working heights varied from pit to pit but in the Kent coal fields the height was usually around four feet. At the bottom of a steep slope known as 2/20s drift the tubs were separated into runs of five each and sent on to the pit bottom and thence to the surface. The roof of the tunnel leading to the coal face was supported by six foot high props that held flat eight foot props that had been sliced in two. The main pump was called The Lady Grey and she pumped 2,000 gallons a minute to the surface. Throughout the workings were smaller pumps. Once the water had been pumped to the surface it ran in channels to large pits connected by gullies. When the sediments of coal dust and iron ore had settled it was possible to swim in the water. Even with the pumps working all out some areas still became water-logged and would have to be shut down temporarily

In the 1920s Tilden Smith instigated a drop in pay. A certain amount of the miners' wages would be held back each week, the idea being that when the pit was shown to be in profit the amount accrued would be paid back to the miners. In those days, of course all records were kept in ledgers and entered by hand. At the end of the year when the back pay was due the ledgers could not be found! The union claimed they had been destroyed on purpose. Whatever the truth the miners all lost out and never received their back pay. In 1921 the management at Tilmanstone Colliery issued a notice about pay. It laid out the pay structure at the colliery. Surface workers earned less than those at the coal face, and those underground still earned varying amounts because they were paid by the number of tubs and the quality of coal they obtained. Very often surface workers were older men who could no longer work underground or men who had suffered injuries that prevented them working at the coal face or undertaking any heavy manual labour. In 1927 a miner was earning about £1.4.6d for four days work. In 1931 the final pay for a full week was still less than £4 a week. During that time deductions were made from a miners' pay for such items as pit props and having tools sharpened.

*Pillar and stall
The pillar and stall method of mining was widely used in the early years of the nineteenth century, but it was a method that had been used in the earliest days of coal mining. It involved cutting the coal into rectangular blocks or pillars that were then cut down into smaller pieces, a method known as 'robbing the pillar'. This was a dangerous procedure because when the last pillar was removed the roof would subside.

*Longwall
In this method the aim was to clear the whole seam in a continuous operation. Longwall advancing was working away from the shaft towards the boundary. This required road maintenance to keep the road clear until the coal was totally removed. Longwall retreating involved working from the boundary back towards the shaft, the roads being driven deep into the coal then abandoned as the coal was removed.

Tilmanstone Colliery

Tilmanstone was always considered a safe pit and pillar and stall working was less arduous than long wall working that was employed in other pits. However, in 1931 a string of tubs broke loose and derailed resulting in three men being trapped. Of the three men one was uninjured and was able to protect the others from falling debris. Prompt efforts of shoring up the roof permitted the successful rescue of all three men. Sydney Padfield who had protected the injured miners was awarded the Edward Medal for Courage in June 1931 by King George V at Buckingham Palace. That was the first of such awards to go to the Kent coal fields. Apart from the constant threat of water seepage fresh air was another great concern. Oil safety lamps were used daily to test the air. The lamps had a rubber bulb attached; when this was filled with air and squeezed the air that flowed across the flame would reveal any dangerous or noxious fumes by changing colour. This sounds very primitive now but it seemed to work.

The miners worked 24 hour days in three shifts, 6am until 2pm, 2pm until 10pm and the night shift from 10pm until 6am. They had a twenty minute break. These meal breaks were known as snap, but both the food and the tins in which they were carried all became known as snap. Just as the Cornish fishermen and tin miners had Cornish Pasties so the miners ate Pit Pasties, a portable meat and vegetable dish enclosed in a pastry shell to protect it. The recipe for Pit Pasty is at the end of this section. Tilmanstone like other pits had a medical room with a nurse in attendance and a doctor made weekly visits to each of the Kent pits. Each pit also had its own ambulance. The medical care of miners was probably better than in many other occupations at the time. These medical facilities were only for the miners themselves. Their families, this was before the advent of the NHS, had to pay for any medical care. To help offset these expenses the families paid into a scheme run by the colliery to ensure medical care if needed.

Pit head baths were not installed at Tilmanstone until 1929 and then only at the instigation of the first Labour Government. During the Second World War the village of Elvington saw its share of bombing raids because of its proximity to the port at Dover and the City of Canterbury to the north, both prime targets for the Luftwaffe. The Battle of Britain also was fought in the skies above that part of Kent. Elvington had its own Home Guard, an enthusiastic but badly-equipped group of men. Their armoury consisted of five Canadian rifles from the First World War and five clips of bullets. Many male and female residents served in the armed forces during the war. It was during the war that the idea of nationalising the pits began to draw support from both inside and outside of the mining industry. Along with nationalisation the miners also wanted a nationwide union to represent them regardless of which part of the UK they were working in. In spite of the need for coal to help maintain industry and continue the war effort miners' wages were still low and working conditions bad when compared to other workers. In 1941 the Essential Work Order was imposed. This legislation made each colliery a 'scheduled undertaking'. No miner could be sacked, but neither could any miner chose to leave.

Tilmanstone No.2 rescue team.

In 1944 the National Union of Mineworkers was founded at a conference in Nottingham, and in 1945 the Kent area miners formed their branch of the NUM with an initial membership of 5,100 members. At the end of the war the general election brought the Labour party into power. Their socialist policies included the nationalisation of many industries including coal mines. Tilmanstone, along with the other 970 pits in the country became nationalised on January 1st 1947 (Vesting Day), but it was never considered to be an economic pit with closure being discussed as early as 1967. With the introduction of nationalisation the collieries came under the management umbrella of the National Coal Board (NCB). The NCB flags were raised, bands played and a new era began for British miners. For those men who had worked under the sometimes paternalistic yoke of private ownership nationalisation was greeted with enthusiasm. For supporters of socialism public ownership could only be a good thing, couldn't it? The NCB chairman, Lord Hindley, announced that, *'we are one family now,'* he went on to say that, *'if they all worked hard and worked together they would make nationalisation a great success.'*

Some members of the NUM had been offered seats on the board of the NCB by Emanuel Shinwell, the minister in charge of the nationalisation plans. Wisely they refused and many heated words were exchanged between Shinwell and Sir William Lawther, President of the NUM, who stated that they were not the administrators but were there to represent the workers. Nationalisation was not the boon it had appeared to be for some miners. Before 1947 the number of collieries had stood at 970. By 1957 this number had dropped to 822. By the end of the 1950s the government together with the NCB began to implement a pit closure programme. The great socialist ideals of public ownership were not working out as the miners had envisaged. Working for the government was actually no better than when the mines had been in private hands and many miners were disillusioned and felt that the government had let them down. Another industry that had been nationalised was the railways, which became British Rail. The coal from Tilmanstone was transported via the East Kent Light Railway to be carried onward by the main Dover to London line. Economy of transportation was not 'important' as it had been during Tilden Smith's creation of the aerial ropeway.

By the late sixties there was widespread frustration at the NCB and its tactics, the main issue being disagreements with management over piecework rates. These and a demand for an eight hour day for surface workers sparked an unofficial strike. It began in Yorkshire and within days 130,000 miners were on strike. Following discussions between the TUC General Secretary, Vic Feather and the NCB Chairman, Lord Robens, the NCB agreed that if the strikers returned to work the board would, in a few months introduce the eight hour day for surface workers. Suddenly the miners were aware of the power they held. At the NUM conference in 1971 pay increases were called for. This was dismissed by the NCB working under government guidelines. An overtime ban was called from November 1st 1971. The NUM then went on to campaign and to ballot the membership on strike action; 58.8 % were in favour. The NCB would not talk to the NUM and the outcome was that on January 9th 1972 the entire British coal industry, including the Kent pits, was on strike. After a month with no leeway given on either side and coal stocks becoming depleted the Prime Minister, Edward Heath declared a state of emergency. The government agreed to set up a public inquiry into miners' pay, but the miners refused to return to work before the inquiry was completed. The inquiry, chaired by Lord Wilberforce, concluded that a definite and substantial adjustment in pay was called for in the mining industry. In addition to pay increases a further sixteen concessions were asked for and obtained. The miners returned to work feeling euphoric at their hard-won victory. In the autumn of 1973 miners and their union officials feared that policies put in place by the government would erode their pay. The NUM acted by introducing an overtime ban and the government's swift reaction was to initiate a three day week for industry across the country. Shops, offices and TV stations were all severely restricted in their use of electricity. At the beginning of February 1974 the NUM was making plans to strike when Edward Heath announced the dissolution of parliament with a general election to be held on February 28th. In calling an election at such a sensitive time the Conservatives hoped that the British people would blame the NUM and its members for the fuel and power crisis and re-elect the Conservatives. To have been re-elected would have given the Conservatives more bargaining power to deal with the NUM. The strike went ahead and the Conservatives lost the general election. The Pay Board, which had been asked to look into miners pay, recommended a rise in wages, pay for unsocial hours and other benefits. The miners returned to work on March 11th 1974.

For the first time in many years investment was put into the industry while £200 million was set aside in a scheme to benefit victims of pneumoconiosis. In 1979, however, a Conservative government under the leadership of Margaret Thatcher was elected. By 1981 she had authorised the NCB to close 23 pits, followed in the autumn of 1982 of the disclosure by the miner's leader, Arthur Scargill of the NCB's plans to close between 75 and 95 further collieries within the next ten years. Thus began the turbulent years of the 80s that would see violence, prejudice, social unrest in the collieries and the pit villages in east Kent. Neighbour would be set against neighbour, miners who did not live in pit villages were vilified by people in the towns and villages where they lived, and families would be split over opposing views about the strike.

The climax of the ongoing dispute was reached in the spring of 1984 when on April 19th a Special Delegate Conference called on all NUM members nationwide to support strike action. Support in the Kent pits was strong and long-lasting. In the early days of the strike on November 19th 1984, 95.9% of Kentish miners were on strike and by 1st March 1985 that had only dropped to 93%. This was a much higher figure than comparison percentages from other coalfields. Throughout its sixteen months duration the strike was the direct cause of 11,000 arrests of miners, 7,000 injuries and eleven deaths. In addition 1,000 men were sacked for supporting union policy in what was the most bitter industrial conflict to be seen in the history of the trade union movement.

TILMANSTONE COLLIERY

In spite of legal moves to cripple the NUM in November 1984 it continued to fight for its members' futures into 1985. Following a meeting between the TUC and the NUM in March 1985 a slender majority vote agreed to end the strike and return to work. Miners in Kent and three other areas voted to continue the strike. There was no negotiated settlement reached with the NCB. The strike had cost the British people £12 billion in hard-earned tax money spent by the government in their efforts to destroy the NUM. This was money badly spent in an ill-conceived vendetta against the NUM and its members.

The then chairman of the NCB, Sir Ian McGregor, wrote in his memoirs that at the termination of the strike the Prime Minister, Margaret Thatcher, had been only days away from conceding defeat and giving way to the striking miners. Following the end of the strike in 1985 the NCB changed its name to British Coal, closed 79 pits, caused the loss of 100,000 jobs, and £4 billion over four years. By 2002 there were just nine collieries employing 3,000 people in the UK. Tilmanstone Colliery closed for the final time in 1986. During its eighty year history twenty million tons of coal had been mined there. Now in the first decade of the twenty-first century all trace of its buildings have vanished and the site is home to warehouses and light industrial units, including a giant salad processing plant of 15,000 square feet, that has created over 500 new jobs. The processing plant makes, amongst other things, sandwiches for Marks & Spencer. The final financial commitment for this site by both SEEDA (South East England Development Agency) and English Partnerships was £4.95 million.

Police arriving by coach at Tilmanstone during the strike of 1984.

Police and miners confront each other during strike.

Pike Road still exists but it is no longer the leafy footpath that miners walked along as they went to and from their homes in Elvington to work at Tilmanstone Colliery. Today the area is travelled by large articulated lorries bringing in produce and goods from across the EU and new foreigners are to be seen in the areas of Deal and Dover. Just like the 'foreign' miners of eighty years ago the new migrants have come to Kent to improve their lives and earn money for their families and, just like the miners, their welcome has not always been a warm one. This is sad because the people now raising objections to the migrants are probably the children and grandchildren of miners. Many of today's migrant workers are bused in to work at the industrial units in the area now known as the Pike Road Industrial Estate.

Felling of headgear, March 1987. It was subsequently sold for scrap.

Recipe For Pit Pasties

Ingredients

8 oz self-raising flour
Pinch of salt
4 oz margarine
2 oz water
12 oz boneless beef, cubed
1 medium onion, sliced thinly
1 lb potatoes, peeled and thinly sliced
1 teaspoon dried parsley
Salt and pepper to taste
2 oz of water or as needed

Directions

1— Using first four ingredients make the pastry by rubbing the margarine into the flour until a bread crumb consistency.

2— Mix together with water until a dough is made.

3— Divide dough into 2 equal pieces.

4— Roll out into circles; use a plate if needed to cut to shape.

5— Place half the potato slices in a line down the middle of each circle.

6— Leave a space of about an inch at each end.

7— Lay onion slices over the potatoes.

8— Season with salt, pepper and some parsley.

9— Place half the cubes of meat onto each circle and season again.

10— Fold the edges over and cut three slits into the top.

11— Place the pasties onto a greased baking sheet.

12— Bake for 45 minutes in an oven pre-heated to 200 degrees C (400 degrees F).

13— Remove from oven and reduce heat to 175 degrees C (350 degrees F).

14— Spoon a teaspoon of water into each slit.

15— Return to oven for 15 minutes or until golden brown.

Colliery canteen, Timanstone, 1980s.

Snowdown Colliery

Exploratory bore hole at site of what would become Snowdown Colliery, *circa* 1905.

Just a year later than at Tilmanstone Colliery, construction of the Snowdown Colliery was started in 1908. The first sod was turned on 28th February 1807 by Mrs Weston Plumptre, the wife of the landowner upon which the colliery was to be sited. The colliery was to be under the management of Mr. H P Nicholson ME.

Snowdown in the early 1900s.

Early view showing railway sidings with plate layers working.

The site was fortuitously at the side of the main railway line between Dover and Canterbury which made transportation of the coal much easier. The small hamlet of Ackholt was nearby, lying between the villages of Womenswold and Nonnington. Coal had first been discovered in the area in 1896 at the nearby Fredville boring. The pit was under the management of Arthur Burr's Foncage Syndicate. Initially his miners lived in neighbouring towns, but this was not an ideal situation either for management or the miners. The nearby villages of Womenswold and Nonnington saw the miners as an unwanted invasion among their quiet rural homes and in the nearby town of Dover landladies were reluctant to rent rooms to them. In 1909 shafts Nos. 1 and 2 were started but No. 1 was abandoned at a depth of 190 feet when it met with a natural cavern into which a large amount of water was flowing and 22 men were drowned. No. 2 shaft was more successful, and was used as the haulage and updraft shaft. As a haulage shaft it had a capacity of 80 tons an hour. No. 3 shaft was of the same diameter and also used for haulage and updraft.

November 1912, coal had been reached at 1,370 feet. Arthur Burr is seen in the centre with handkerchief in coat pocket.

The Beresford seam was reported as being of very fine quality. A geologist, Mr. George Hislop, wrote to the directors of the company stating that the Beresford coal seam looked very good indeed. Following the report Snowdown Colliery shares appeared on the London stock exchange. Along with such an optimistic report the shares in companies of which Arthur Burr was a director all rose in value. In November 1912 the first coal was mined at 1,370 feet and brought to the surface thus making it not the oldest but the first commercially viable pit in Kent. The event was recorded in a photo. In January of the next year a coal seam five feet six inches thick enabled 800 tons to be mined per week.

Celebration of raising samples of coal from the Beresford seam, *circa* 1912.

Picture in Beresford seam where the first coal had been taken, February 1913.

Snowdown Colliery

SNOWDOWN COLLIERY

The No. 1 shaft, the Beresford, finally reached a depth of 1,510 feet. Shafts Nos. 1 and 2 were deepened over the next two years to reach the No. 6, (Millyard) seam. No. 2 reached a final depth of 3,011 feet and No. 3 reached 2,952 feet. A major problem as with other Kentish mines was the water. At Snowdown there was a continuous inflow from the chalk and greensand beds above. This resulted in up to one million gallons of water per day being pumped out into the River Stour.

At the start of 1919 the new chimney was nearing completion and at 150 feet high it stood 30 feet taller than the old one. In 1920 the Emergency Powers Bill was passed by Parliament. This increased wages for a period of six months, but at the end of the six months the owners drastically reduced wages and the miners at Snowdown went on strike as a result. Notices were posted at the pit announcing the reduced rates of pay. The owners hoped the colliery could re-open on June 13th and a number of men wished to return, but they were prevented from doing so by a strong force of pickets. About 300 pickets gathered in the road beside the train station. The company went into deeper financial crisis and eventually it went into receivership.

Although no mining operations were being carried out at the pit, pumping operations continued so that the colliery would not flood and could be sold as a going concern. Snowdown was subsequently purchased by Pearson & Dorman Long, in 1924. Dorman Long was an established steel manufacturer in the north east of England, who had also started a new colliery in the Kent coalfield at Betteshanger. Following the purchase the colliery underwent a major modernisation programme in which the original steam-powered winding plant was scrapped and replaced with an electric one. A site covering 600 acres was then bought and a Public Utility Society, Aylesham Tenants Ltd. built the village of Aylesham to accommodate 650 families.

In May 1926 the General Strike brought the country to a standstill. The strike lasted for nine days but many of the miners stayed out until November. In September of the same year work got under way to build the new town that would house the miners and their families. The first stage was to comprise 400 houses; half were to be built of traditional materials while the other half were to be of concrete and steel. In spite of the miners being on strike when the building began by the end of September over 200 miners had returned to work and produced 60 tons of coal in one shift. Bricklayers and various other building workers walked to the site each day, often from Dover, a journey of several miles, glad to have found work during a period of high unemployment.

In 1928 the owners again attempted to reduce wages. Attempts were made to re-open the colliery but only 96 men showed up for work. By the time the early shift finished a large crowd had gathered outside the pit. There was a heavy police presence and one miner was injured. The following day just 40 men showed up at the pit gates protected by 70 policemen who escorted them from the colliery to the railway station. The risk of a further strike at the beginning of October 1929 was over the rates of pay in Kent pits. At that time there were no uniform pay rates and each colliery owner could pay however much he wanted; Kent was the only coalfield where there was no district agreement about pay rates. The Kent Conciliation Board failed to reach an agreement with the Kent Mine Workers Association (KMWA) and the mine owners, but a ballot went against strike action at that time.

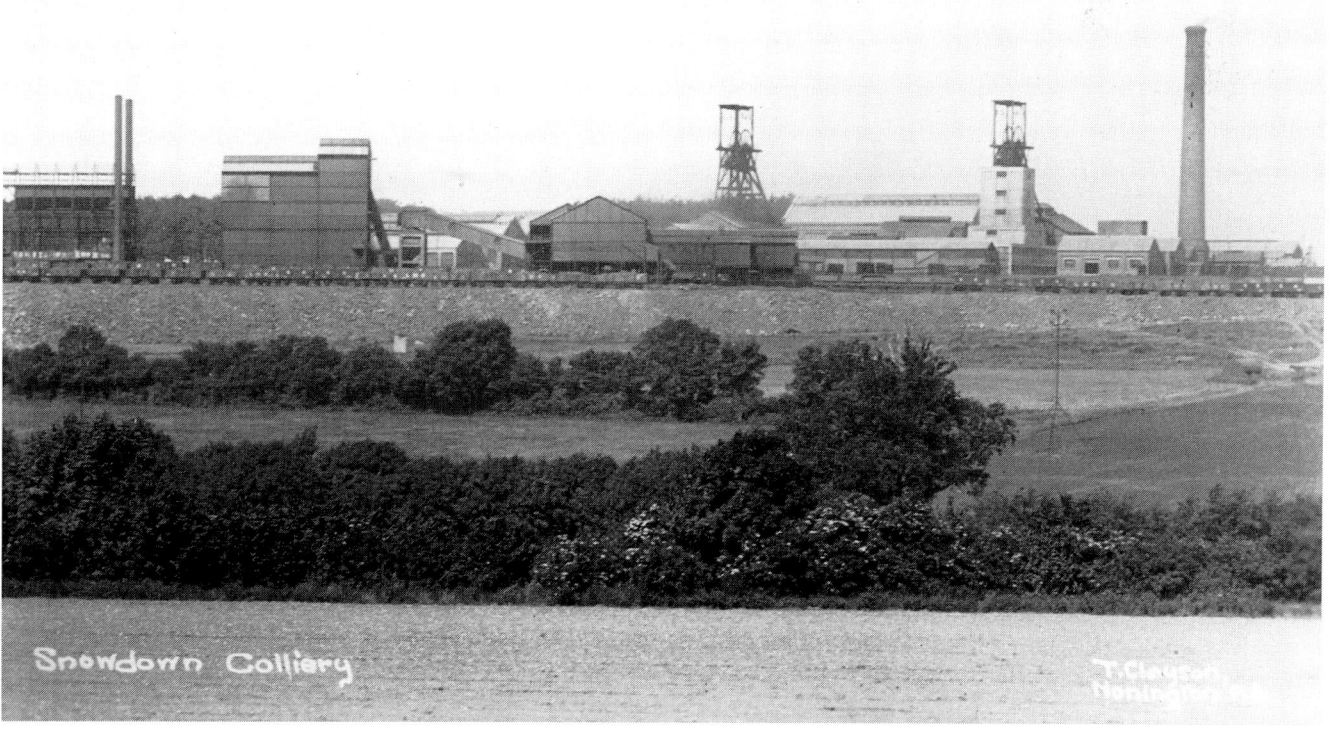

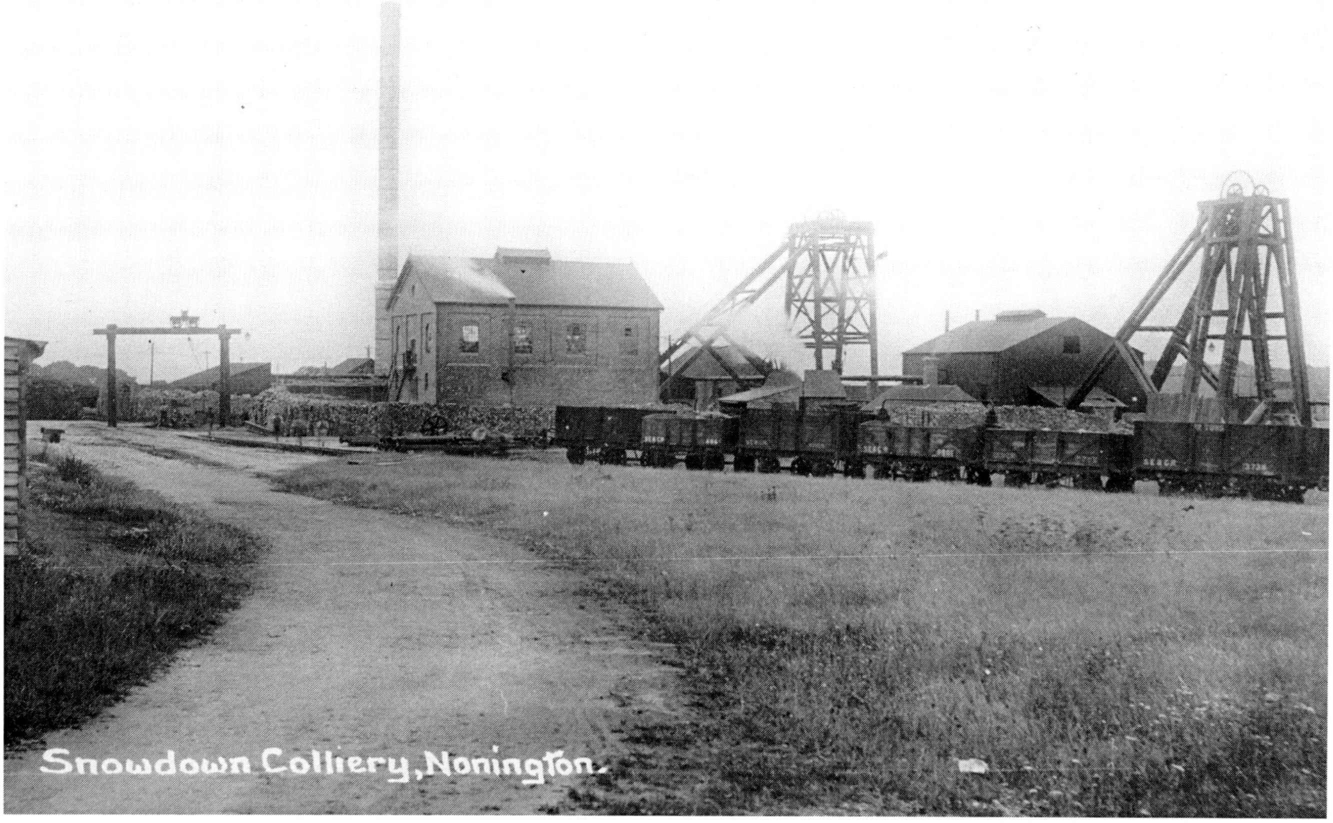
Snowdown Colliery, Nonington.

Snowdown Colliery
T. Clayson,

SNOWDOWN COLLIERY

Following the General Strike miners arrived at Snowdown from many northern pits where they had been blacklisted. The conditions they found in Kentish collieries were among the worst in the country with many old-fashioned working practices still accepted as the norm. Because of such high unemployment throughout the 1930s these conditions were tolerated as the only other alternative was to be out of work. With unemployment at such a high most people were glad to have steady work and a roof over their heads as were the miners of Snowdown, but as always every barrel had its bad apples. In March of 1930 three young children were observed going from house to house in the village of Adisham. When a policeman arrived on the scene he saw that the children were carrying a letter saying that there was no food in their house, and that their father was sick, the letter signed by the children's mother. The baskets the children were carrying contained bread, cheese, tea and a pair of boy's boots. The policeman took them home to Aylesham where he spoke to the parents who admitted sending the children out begging.

On March 7th, in court, a doctor testified that the father was fit for work but had only worked one and a half shifts at Snowdown Colliery in the past month. Other witnesses stated that he was well-known for being lazy. He was sent to gaol for one month. In January of the same year a letter appeared in the *Dover Express* complaining of the poverty being suffered by Kent miners due to high rents and low wages. The writer went on to say that when jobs became more plentiful again the miners would be able to take their pick and move away from Kent.

Other crimes committed during that period have a ring of familiarity about them; a refreshment hut at Aylesham was broken in to and goods taken while at Aylesham Halt, the nearby train station, the station master's office was burgled with £2. 1s. 3d. in cash being stolen.

In April, 1930 a low temperature fuel carbonisation plant was installed. This proved a valuable addition as it was capable of extracting oils from the coal while the residue was still a good burning household coal.

Aerial view of Snowdown, 1931.

Until 1932 when union pressure forced a change wages were paid on what was known as the 'butty' system. This meant that the chargeman, a man over a group of miners, would be paid a lump sum for their work and the chargeman then paid the miners. It was standard practice for him to keep any 'odd' change for himself. If a miner was due for example 10s. 11d. the chargeman would keep the 11d. Management turned a blind eye to the practice and if a miner complained he would be moved to a less profitable working 'stall' or area.

Of all the Kent collieries Snowdown was the deepest. It was also the most humid and hottest. Its nickname was Dante's Inferno. Many miners regarded it as the worst pit to work at in the whole of the UK and most Snowdown miners worked naked because clothing became too uncomfortable in such conditions. To combat dehydration the miners would drink up to 24 pints of water in an eight hour shift. Heat exhaustion was a frequent hazard for the men. In March 1935 Snowdown opened its first pit head baths, built by the Miners' Welfare Committee. A plaque on the building recorded the achievement. It was a great benefit for the miners, and also for their wives who would no longer have to deal with their husbands arriving home covered in coal dust.

The miners who came to work at Snowdown Colliery were mainly from South Wales, Scotland and the northeast. Living at Aylesham the village developed in isolation from the surrounding Kentish communities. A busy person around Aylesham in the early days was Nurse Harrison, the local district midwife. The majority of births were at home and Nurse Harrison and her bicycle were regular visitors to the village.

The residents formed various sports clubs such as football, bowls, cricket etc. and these are still active at the present time. One sport not often seen around the British Isles was well supported in Aylesham. During the 1950s a great number of US airmen were stationed at the nearby base of Manston. It was with the help and enthusiasm of some of them that a baseball team was formed at Aylesham. The Americans helped in obtaining uniforms and teaching the miners how to play. Baseball was played enthusiastically for many seasons.

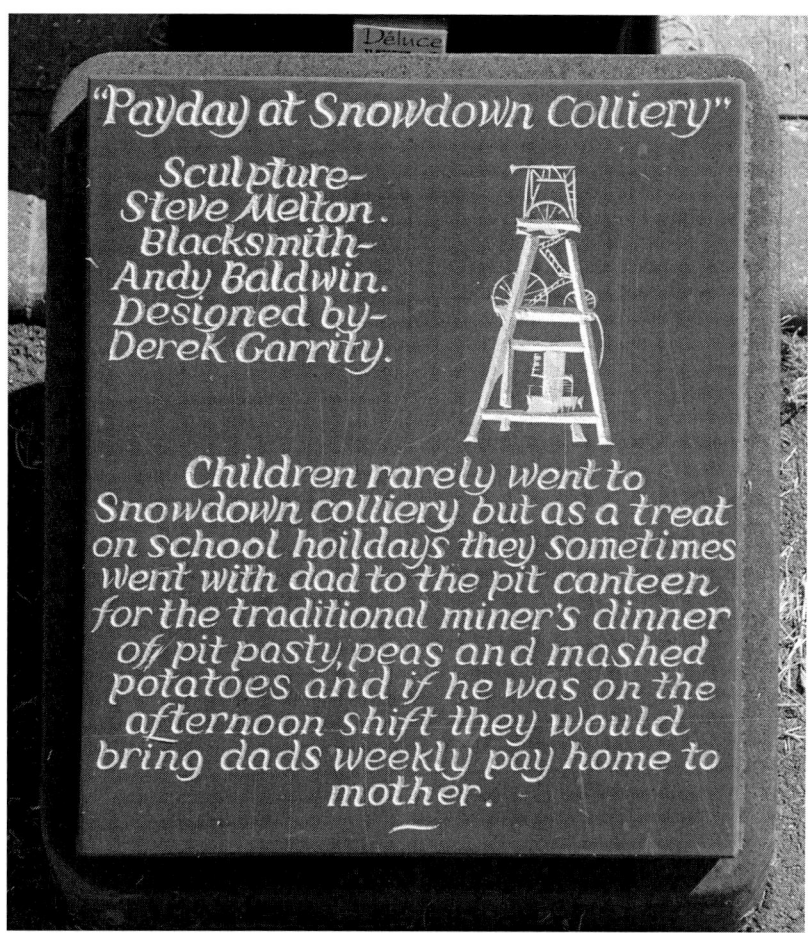

'Pay Day' statuary at Aylesham.

Snowdown Colliery

The residents of Aylesham formed a brass band as early as 1926. Known as the Snowdown Colliery Welfare Band its musicians consisted entirely of miners. Three years later a male voice choir came into being. Today the band is still playing and performs at many venues during the course of a year. The band has around 30 members and although only one is an ex-miner several of the band member's fathers were miners.

In 1947, along with all the other coal mines in Kent, Snowdown became part of a nationalised industry. One reason given for the advantages of nationalisation was improved safety. Pits, when in private ownership generally had an appalling safety record. That did not mean just safe working conditions as far as preventing accidents but the gross lack of continuing care for disabled miners by private owners. Under the NCB in 1952 there began regular medical examinations of all miners, but by 1955 deaths from pneumoconiosis still topped 2,000 men industry wide. It is not a quick and easy death; the results can still be seen around east Kent as ex-miners continue to suffer from this industrial disease of the lungs.

Kentish coal had always been some of the most difficult to mine and because of that, also the most expensive. Under the control of the NCB plans were soon under way to begin the closure of Kentish pits as early as 1960. In the late 1950s and early 1960s nearby Richborough Power Station went on line; it was coal powered and most of the coal came from the nearby pits. Between 1962 and 1971 over 3 million tons of coal were used. In 1971 the power station made a bad choice and switched to being oil-powered. The increase in oil prices eventually made that unviable. In 1989 Orimulsion, a liquid made from bitumen and water and imported from Venezuela, began to be used instead.

In the early 1970s mine closures had already begun. In 1964 Britain had had 550 pits; six years later that number had dropped to 250. In 1972 the miners' union balloted for a strike and on January 9th 1972 all the pits were on strike, the first time there had been a national strike in the mining industry since 1926. This strike was about pay. The Prime Minister, Edward Heath, declared a state of emergency and instituted a three day working week to save power. On February 19th agreement was agreed between the NUM and the government and miners resumed work on February 25th. Because of the incomes policies of the then Conservative government the NUM introduced an overtime ban in the autumn of 1973 in pursuit of a pay claim. A further strike followed in 1974 that resulted in the Prime Minister dissolving Parliament prior to a General Election. The result of a Pay Board enquiry was not made public until after the election. The Board's findings were accepted and the NUM, together with the newly elected Labour government, held talks that thrashed out a real plan for the future of the coal industry. For the time being!

Arthur Scargill, president of the NUM, visits Snowdown in November 1983.

By 1975 there were less than 1,000 miners working at Snowdown and total production, along with Tilmanstone and Betteshanger was around a million tons a year. Most of the coal was used as a coking blend for use by steel producers, another British industry that was in the doldrums. At the start of the 1980s the British government and the NCB saw that the only way to make the coal industry viable was to close 'uneconomic' pits. This was disputed by both the miners and the NUM who were convinced that lack of investment and bad management was slowing the industry from reaching its full potential. In 1983 two exploratory drifts were built to the No. 7 seam approximately 1,312 feet below No.6. However, it was thought to be uneconomic to develop and Snowdown Colliery became the NCB's first target pit for closure in Kent.

In the early 1980s trouble reared its ugly head and by 1984 the NUM was running a strike that would ultimately see the closure of all the Kentish coal mines and a loss of that sense of community that had long been the byword of pit villages throughout the area. The strike set friend against friend and families against their own relatives.

The use of Orimulsion at Richborough Power Station caused many environmental problems, its use ending with a lawsuit by a local farmer being settled privately. The cost of coal versus Orimulsion was about the same, at 0.5p per kilowatt hour, but there was no going back, with the local coal no longer available and the power station was closed in 1996.

Richborough Power Station being demolished, 2008.

Snowdown Colliery

As early as 1982 a document had been leaked from the NCB to the NUM that listed the 75 to 95 pits that were due to be closed over the next ten years. In January 1983 the Kent NUM opposed a deal by the unions and bosses regarding redundancies at Snowdown.

Strike action was overturned by NUM officials. Among those listed for closure was Snowdown. It was that leaked list which helped to spark the industrial action that resulted in the disastrous strike of 1984-85, a strike that was to last for sixteen hard and violent months. The length of the strike caused great hardship and indeed physical and mental illness among the mining communities. During the strike, nationwide there were 11,000 miners arrested, 7,000 suffered injury and eleven people lost their lives. It was said that the police in their role of keeping the peace and upholding the law were more like the Conservative government's private army than a police force.

In order to save the union the NUM executive voted by a small margin to return to work. The only coalfield to vote against this decision was Kent. On March 3rd 1985 without anything being gained and a great deal having been lost the miners returned to work. The strike was a controversial and bitter dispute with much harm done to people on all sides. The Conservative government fought the miners with every tool at their command. They had the manpower, in the guise of the police and the money in the form of collected taxes to pursue the striking miners. Although it was not really acknowledged by the media, who for the most part supported the government, it was the taxpayers of Great Britain who paid for the police overtime needed to carry out what looked very much like a private war between the government and the union. The attitude of the government, especially that of Prime Minister Margaret Thatcher appeared to be one of vindictiveness.

Worker at the coal face, 1985.

Upon the miners return to work in 1985 the men employed at Snowdown felt that under the reorganisation of the NCB, now called British Coal, the writing was on the wall and that the Kent pits had only a short time to go before they were closed. Indeed, Snowdown which in its heyday had employed 3,500 men, closed in 1987. Since its closure nothing, apart from capping off the shafts, has been done at the site. The buildings stand empty among weed-strewn pathways and the rusting main gates hang precariously on their hinges. The only life encountered at this once busy, noisy and productive colliery today are birds who nest within the deteriorating buildings and the ubiquitous foxes who have made their homes in this quiet and deserted complex.

Across the road amid its overgrown car park stands the disused Snowdown Working Men's Club. Some of the houses of Aylesham Village have been taken over by the local authority and others are now privately owned. It is a quiet and for the most part neat and tidy village. Many of the houses at Aylesham look out over the green and gently rolling Kentish landscape, the residents benefiting from views that many people would pay money to live with, but the coal, still left under that same countryside meant a great deal to the past generations who lived there. They fought for it and worked hard to maintain a way of life when they knew of no other.

It was pointed out by the chief executive of Dover District Council just prior to the closure of Snowdown Colliery that *there are now more commercial vineyards in Kent than pits*. Since the colliery closed the villagers of Aylesham have not spent time bewailing their loss. By January 1998 the Aylesham Neighbourhood Project was up and running, its aim being to encourage action and growth within the village that would be of direct benefit to the residents. The building that previously housed the local school gym has been redeveloped to house a pre-school for local children that offers summer play schemes, after-school clubs, homework clubs and parents' groups. The middle section of the school is now a well-stocked and welcoming public library. In other parts of the building training schemes and support from agencies that have been assisting the Single Regeneration Scheme, (SRB) are on offer. A telecentre equipped with fifteen computers is available to help both school children and adults to gain computer literacy. Around the building ten workshops have been established and rented to small businesses and this too has been a success.

Main Gates, Snowdown, 2008.

Across the car park from the library the main school buildings have also taken on a new identity. Refurbished and well-equipped the rooms there now earn their way by being rented out as conference rooms. Aylesham's proximity to both Dover and Canterbury and the village being just off the A2 with free parking, makes it a popular venue for companies to hold meetings and seminars. On the ground floor of the building is the Sunshine Café, offering snacks and hot home-made meals at reasonable prices for both local residents and those attending meetings in the building. It is run by a local lady who is a qualified chef with catering experience. A catering service is also offered for weddings and birthday parties.

A life-sized statue of a miner with a small child on his shoulder and a little girl walking beside him stands just nearby the ex-school complex. The figures stand in front of a row of small coal cars on rails. The statue is called Pay Day. Traditionally on pay day children would go to the pit canteen and their fathers would give them the week's pay to carry home to their mothers. It is a lovely and poignant reminder of how things used to be. In the centre of the village, set on a concrete plinth is a winding wheel from the colliery. It was unveiled following the closure of Snowdown Colliery in 1987, together with a commemorative plaque.

A retired ex-miner who had followed his father into the pit at the age of fourteen was proud of his long association with the colliery and still lived in a house in Aylesham Village. He was even more proud of his children, both of whom had attended grammar school and gone on to be university graduates. One works in the field of medicine while the other follows a career in the legal profession.

Because the other colliery sites of east Kent have been almost totally demolished two of the remaining buildings at Snowdown have recently been designated Grade 2 listed by English Heritage. A study is being carried out by The Snowdown and Kent Coalfield Heritage Group, (SkaCH) a group formed of local residents. The study will explore the continuing use of the colliery site with the objectives:

The restoration and future development of the site.
To provide a centre for the celebration of the Kent Coalfield.
To create activity and employment.
To offer opportunities for education and training for local residents.
To develop opportunities for recreation and enjoyment for both the local and wider communities.

With these aims and challenges in the future it should be that the coalfield of east Kent will be a part of the lives of the people of the county for many years to come.

Kent's Last Days of Colliery Steam

by Tom Heavyside

Introduction

When the coal industry was vested in the state on 1 January 1947, included among the myriad assets in Kent inherited by the newly-created National Coal Board were nine steam locomotives. They were based at the collieries at Betteshanger (4), Chislet (2) and Snowdown (3). The fourth working colliery in the county, Tilmanstone, was originally shunted by locomotives belonging to the East Kent Light Railway, British Railways carrying on the tradition when the railways were nationalised in January 1948.

At the start of 1947 both Betteshanger and Snowdown relied entirely on six-coupled locomotives, three side tanks and four saddle tanks, for their internal requirements. The three side tanks had emanated from the Leeds-based company Hudswell Clarke in 1918, the same company being responsible for one of the saddle tanks in 1923, with the remaining three being constructed by the Avonside Engine Company of Bristol in 1927 (2) and 1931. These seven locomotives were previously owned by Pearson & Dorman Long and, except for 'St Martin', carried the name of a former Archbishop of Canterbury. (The small parish church not far from Canterbury Cathedral, and reputedly the oldest in the Anglican Communion, is dedicated to Saint Martin). At Chislet there were two 0-4-0STs previously owned by Chislet Colliery Ltd, one built by Peckett of Bristol in 1924, the other by W.G.Bagnall of Stafford in 1925.

Over the ensuing years the NCB purchased three new steam locomotives for use at its Kent collieries, a Bagnall 0-4-0ST in 1950 for Chislet and two of the sturdy Hunslet 'Austerity' 0-6-0STs in 1954 for Betteshanger. In addition a pair of Yorkshire Engine Company (Meadow Hall Works, Sheffield) 0-6-0STs were bought second-hand from steel manufacturers Samuel Fox & Co Ltd of Stocksbridge, near Sheffield, in 1959, as was a Peckett 0-4-0ST from the famed chocolate makers Cadbury of Bournville in 1963. The final acquisition was an 'Austerity' 0-6-0ST transferred from the NCB Holditch Colliery at Chesterton, near Newcastle-under-Lyme in Staffordshire in 1966. The last mentioned had been built by Hudswell Clarke in 1943 as part of an order for the Ministry of Supply. The new arrivals enabled some of the worn out locomotives to be withdrawn and sold for scrap.

At the turn of the decade into the 1970s there were seven 0-6-0STs listed on the books of the collieries in the Kent coalfield, the aforementioned three Avonsides at Snowdown, the three 'Austerities' at Betteshanger, and Yorkshire Engine Company works No. 2498, built in 1951, at Chislet. By this time some of the work had been taken over by diesel locomotives, with one example being based at each of the four collieries. By the start of 1973 the surviving steam was concentrated at Snowdown, namely the trio of Avonsides that had all started their careers there and 'Austerity' Hunslet works No. 3825 transferred from Betteshanger at the end of the previous year. The other two 'Austerities' formerly at Betteshanger had by this time both been scrapped, while the Yorkshire Engine Company 0-6-0ST at Chislet was a little more fortunate in that it was saved for preservation. The latter was despatched to the Buckinghamshire Railway Centre at Quainton Road, near Aylesbury, in January 1970, where it is now identified as 'Chislet' in honour of its old workplace.

During the 1970s Snowdown was still a productive mine, about 850 men drawing their wages from the pit offices each week, with approximately 300,000 tons of mainly good quality coking coal being brought up the shafts annually. Everyday tasks for the locomotives included shunting the yard and transferring wagons to and from the BR exchange sidings by the ex-South Eastern & Chatham Railway London to Dover via Canterbury East line, two normally being required in steam each day. However, resources were often stretched, especially after 1971 when the youngest of the Avonsides, 'St Martin', became unavailable, this being retained merely as a source of spare parts to help keep its two older sisters operable. 'St Martin' was finally scrapped at the end of 1973. An attempt was made to dispense with steam at Snowdown in late spring 1976 when a couple of ex-BR diesel shunters were transferred from Betteshanger, the installation of a rapid loading system at their previous place of employment having rendered them redundant there. The newcomers soon proved themselves somewhat unreliable, in fact one never turned a wheel in revenue earning service at Snowdown, meaning there was little option but to continue with steam until some more dependable diesels could be transferred in from elsewhere. In the event steam lingered on until August 1979, when the fires were dropped for the very last time.

Not surprisingly, during its last years the management at Snowdown received many requests from railway enthusiasts to visit this Kent steam oasis. To satisfy demand two open days were held when rakes of wagons were shunted up and down the yard much to the delight of photographers and onlookers alike. Happily, the three steam locomotives based at Snowdown during the late 1970s all escaped the breakers' hammer and are still extant. Appropriately the two Avonsides both reside not too far away from their old home, 'St Dunstan' on the East Kent Light Railway at Shepherdswell and 'St Thomas' at the Dover Transport Museum, while the 'Austerity' is currently at the West Coast Railway Company headquarters at Carnforth, Lancashire.

In completing this section I would like to acknowledge the help received from Paul Abell and Bob Darvill. My sincere thanks are also due to the National Coal Board who allowed me free rein of the yard at Snowdown Colliery for a couple of days in September 1974.

KENT'S LAST DAYS OF COLLIERY STEAM

Above: Two of the engines based at Snowdown Colliery are prepared for work outside the two-road, brick-built shed on Monday 16 September 1974. On the left is Avonside 0-6-0ST works No. 2004, 'St Dunstan', built in 1927, while on the right the driver of Hunslet 'Austerity' 0-6-0ST works No. 3825 of 1954, NCB No. 9, keeps a watchful eye on the water level as the 1,200 gallon-capacity saddle tank is topped-up. A close examination of the metal plating below the smokebox door of 'St Dunstan' indicates some patching may be necessary before too long. Note, too, the shunters' poles carried just above the bufferbeam of each locomotive. All the photographs at Snowdown were taken on this or the following day.

Left: The driver of 'St Dunstan' crouches while pouring some very necessary oil into one of the many lubrication points, as he systematically works his way round the motion. Nearest the camera part of the right-hand cylinder casing is visible. The insides of the cylinders measured 15in. diameter with a 20in. stroke, while the wheels stood 3ft 6in. above the rails. Leading from the cab, parallel with the running plate, is the reversing lever. Strong boots and overalls were obviously advisable in this far from sanitised environment, although not always observed!

Right: The driver has now moved to the left-hand side of 'St Dunstan' to attend to one of the lubrication points beneath the boiler barrel. In his left hand is a wad of cotton waste for wiping any excess oil. Another vital job, especially during wet weather or if the rails were greasy when the engine would be more prone to slipping, was to ensure the sandboxes above the leading wheels had an adequate supply available for the day ahead. The brazier at the driver's back stands ready to protect the water supply during times of frosty weather.

Below: On what was a pleasant cloudless morning, the driver on this shift appears to prefer a rather natty jumper for footplate work. Clearly evident, attached to the cab side sheet of 'St Dunstan', is the cast oval plate detailing the Avonside works number and year of manufacture, as is the nameplate fastened to the saddle tank. Saint Dunstan, a Benedictine Monk, was Bishop of Worcester and then of London before being consecrated the 25th Archbishop of Canterbury in 960. He is the patron saint of bell-ringers, metalworkers and locksmiths.

Here 'St Dunstan' slowly drags some empty mineral wagons towards the back of the coal preparation plant – a seemingly complex structure. The engine sported a light blue livery lined in yellow, with opposing diagonal yellow and black stripes adorning the front (except for the saddle tank) and, unusually, also on the sides of the bunker. The contrasting coloured stripes were for safety reasons and they certainly made the engine stand out in what could fairly be described as hazardous working conditions, especially for the unwary.

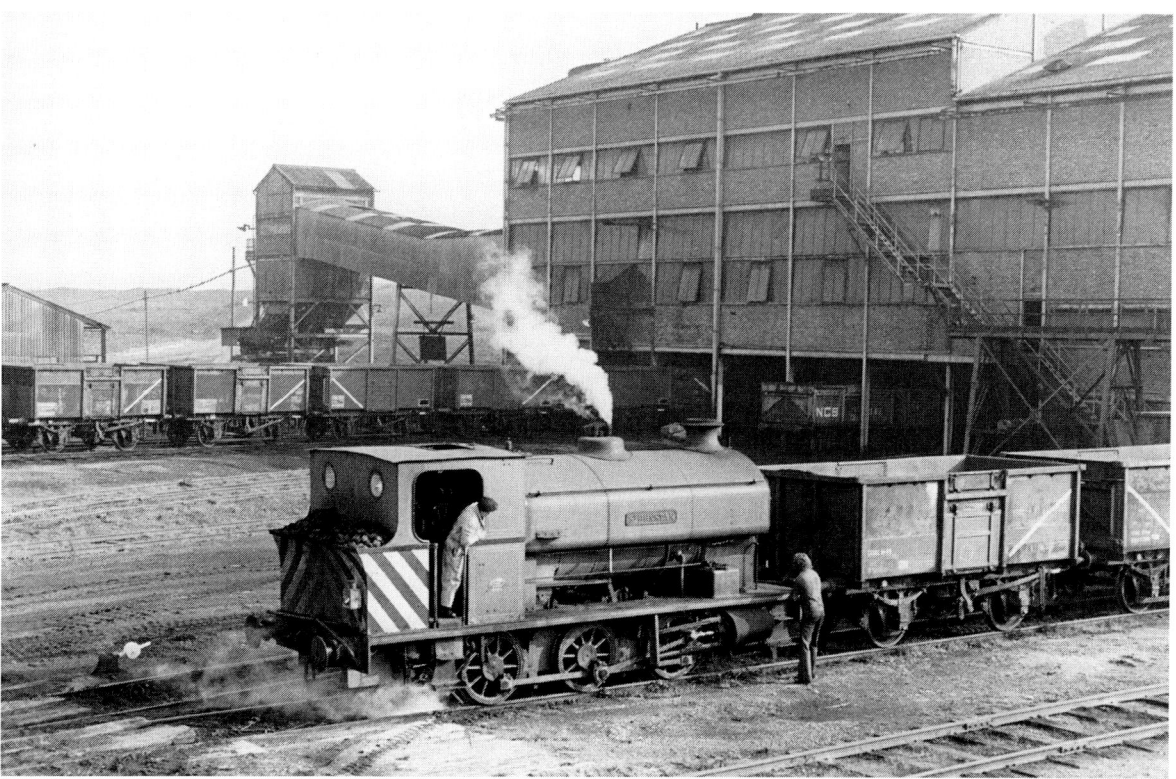

Having reached the other side of the screens and come to a halt, the driver watches his mate closely as he unhooks 'St Dunstan' from the leading wagon. From this angle the 'wasp' stripes can be seen stretched across the back of the bunker. There appears to be plenty of fuel available for the job in hand. The wagon partly hidden under the screens marked 'NCB', was for use only within the confines of the colliery.

A panoramic view of the coal preparation plant as No. 9 draws a rake of fully-laden wagons towards the photographer. A conveyor enclosed within the gantry on the left carried the raw material to the plant, while that on the far right (partly hidden by the trees) was used to transfer waste to the tip. The circular building on the right is the froth floatation building where the coal dust was separated. The 'Austerity' class was designed by the Hunslet Engine Company of Leeds in 1942 at the request of the Ministry of Supply, a simple, rugged, easy to maintain shunting locomotive being urgently required to assist the war effort both at home and overseas. The doyen of the class made its appearance in January 1943. They had 18in. x 26in. inside cylinders, 4ft 3in. diameter wheels set to a rigid wheelbase of 11ft 0in. enabling them to negotiate curves with a minimum radius of 180ft. The boilers were pressed to 170lb. per square inch, providing a tractive effort of 23,870lb. at 85% of maximum boiler pressure, although the Hunslet catalogue stated 21,060lb. at 75% pressure in accordance with their standard practice. In full working order they turned the scales at 48 tons 4cwt. and it was recommended that they ran on rails of not less than 80lb. per yard. They were designed to haul a trailing load of 1,120 tons along a level section of track, 565 tons up an incline of 1-in-100 and 320 tons up a gradient as steep as 1-in-50. Such was the success of these locomotives that production continued long after the war had ended, the 485th and last example not taking to the rails until 1964. Seven private locomotive companies participated in their construction. They were ideal for the harsh conditions experienced at most NCB establishments, over 250 being owned at one time or another by the NCB, most purchased second-hand, some third-hand, together with a further seventy-seven bought new. No. 9 was the last survivor of the three 'Austerities' employed in the Kent coalfield.

Framed by the headgear above the two shafts, No. 9 handles with ease a long link of wagons. Like its two companions, No. 9 was painted light blue but with white lining, although without the warning stripes at the front or rear. Barely visible under a thick layer of grime, boldly stencilled on the side of the saddle tank was the legend 'National Coal Board South Eastern Division No. 9'. This inscription related to the days before the NCB abolished the old divisions in favour of a new management structure inaugurated on 26 March 1967, the collieries in Kent then being linked directly to the NCB headquarters at Hobart House, Grosvenor Place, London, for administrative purposes. From 1 December 1975 the isolated Kent coalfield became part of the South Midlands Area, whose offices where over 150 miles away at Coleorton, Leicestershire.

'St Dunstan' bides its time in the stockyard as a mobile grab crane empties the contents of one of the wagons. The wagons displaying a large 'X' on the sides were again for internal use only. When the author last visited the site in May 2006, even though this was almost twenty years after the colliery had closed, many of the buildings seen here were still standing, although most appeared in a rather woebegone state. However, there was no sign of the headgear that was such a dominant feature of the surroundings, for this had been dismantled in 1988, one of the winding wheels subsequently being erected in the village of Aylesham, one mile north of Snowdown where many of the former miners lived. A plaque reads 'In memory of everyone associated with Snowdown Colliery, their wives and families, 1907-1987' (see page 54).

The weather was a little murky as 'St Dunstan' and No. 9 noisily approached the BR exchange sidings on the morning of 16 September. The main line is to the left of the brick-built, flat-topped signal box, the design dating from 1953 when the original was rebuilt. Access to the exchange sidings was controlled by means of an 18-lever Stevens frame. With windows overlooking all four sides, the man in charge of the signal box was able to keep a close watch on all that was happening around his domain.

During a spell of hazy sunshine, No. 9 again enters the exchange sidings, this time working solo. The proximity of the colliery yard can be gauged from the buildings on the right. The wires erected above the rails on the left, carrying current at 750V dc, were for the use of BR Southern Region Class 71 electric locomotives. When running on the main line the 71s collected power from an electrified third rail, but this method would have been impractical, as well as downright dangerous, in colliery sidings such as at Snowdown, hence they were fitted with dual equipment. The changeover from one system to the other was effected on a short length of track by the signal box where current could be taken either from a third rail or the wire suspended overhead, locomotives halting briefly while they raised or lowered their pantograph.

No. 9 and 'St Dunstan' stand near the BR main line as a Southern Region electric multiple-unit dashes by forming the 10.14 Dover to London Victoria service. Preparing to enter the sidings in order to collect some outgoing traffic is BR Class 33 diesel-electric No. 33046. Note the various point levers. The station at Snowdown (the platforms were only just out of sight beyond the signal box) was originally opened in 1914 for the convenience of the local miners. From May 1969 it became known as Snowdown & Nonington before reverting to plain Snowdown eleven years later in May 1980.

At the end of the shift, flanked by an accumulation of ash raked out from the fireboxes and smokeboxes, 'St Dunstan' and No. 9 pose outside the shed. Behind 'St Dunstan' is an old coal wagon commandeered for use as a coaling stage, the front end having been removed to assist shovelling the fuel into the bunkers. To be glimpsed inside the shed doorway is the third steam locomotive available to the yard foreman at Snowdown in the mid-1970s, Avonside 0-6-0ST works No. 1971, 'St Thomas', built in 1927. Like 'St Dunstan', 'St Thomas' never worked anywhere else other than at Snowdown. It was destined to have the dubious honour of being the last steam locomotive to be used on revenue earning duties in the Kent coalfield, the finale taking place in August 1979. The engine was named after Thomas Becket, the 40th Archbishop of Canterbury, holder of this ecclesiastical office from 1162 until his untimely death in 1170. He was canonised Saint Thomas of Canterbury in 1173. Directly behind the shed is the unclassified minor road leading through the village, while Snowdown station is just to the left of the picture.

A nocturnal view of No. 9 safely ensconced within the confines of the shed for the night. On the right is the ironwork supporting the twin water tanks (see previous picture), while on the left is the wagon that doubled as a coaling stage. There was only room for two engines under cover so, on this particular evening, it was 'St Dunstan' that was forced to brave the elements outside. Such scenes of working steam, playing a vital role in the movement of the black diamonds as depicted in these few pages, can never be replicated. Neither will they ever be forgotten by those who had the opportunity, nay the privilege, to witness just a small slice of what was once a daily drama at first hand before it was all consigned to history. Meanwhile the spirit of steam at Snowdown lives on in the shape of the three locomotives that ended their commercial working days there at the end of the 1970s, albeit, as detailed in the introduction to this section, in what many would consider much more congenial environments.

CHISLET COLLIERY

In the early years of the twentieth century German industrialists took an interest in the embryonic coal industry in Kent and centred their search around the city of Canterbury. In 1911 the Anglo-Westphalian Coal Syndicate Ltd. leased some land near the village of Chislet. The company planned to transport the coal by barge via the Sarre Penn stream and the ancient Wantsum Channel, the waterway that once divided The Isle of Thanet from mainland Kent. Fortunately, because it would have caused the destruction of much natural countryside and archaeological sites, permission for this was refused. Exploration was then directed to an area at Westbere where the coal could be moved by road and rail. As with other sites in the Kent coalfield water proved an obstacle. Coal was reached at a depth of 1,350 feet.

Chislet Colliery was born in 1914. The local director of the colliery was Herr Willi Perits. It is stated in some documents that Herr Perits became a 'Guest of the Nation' at Alexandra Palace in London. This may have been another way of saying that he was interned as an enemy of the country, as Alexandra Palace was used to hold internees during the First World War. The colliery was located on the south side of the main Canterbury to Margate road and was initially owned by the Anglo-Westphalian Kent Coalfield Ltd. At such a politically sensitive time there were questions asked in Parliament about the name and its obvious German connections. The name was subsequently changed to North Kent Coalfield Ltd. and later it became Chislet Colliery Ltd. (Chislet being the name of a nearby village).

Early view of bore holes being drilled at Chitty (nr. Chislet), *circa* **1900.**

It was the most northerly colliery in Kent. The first coal was mined in 1918; two shafts were sunk, the north at 1,470 feet and the south at 1,457 feet. In the 1920s saleable quantities of coal were mined on the site but strikes and lack of repairs by a cash-strapped company caused areas of the mine to collapse. In 1929 the company appointed E.O.Forster Brown, a mining engineer, to reorganise the colliery and a Charles Clark was also employed to improve relations between the company and its employees. These actions were successful and output increased, along with profits, which also improved the social lives of the miners.

Early days of Chislet, *circa* 1920s.

As with other pits in the Kent coalfields the indigenous local people were not experienced in mining and the work force came from traditional coal mining areas of the UK. The largest contingent to work at Chislet came from South Wales, the reason being that some Welsh coal companies held shares in the company developing Chislet. In the early days at Chislet many of the miners lived on the Isle of Thanet in either Margate or Ramsgate where the company had taken leases on several hundred houses. They were brought to the colliery by specially run miners' trains that stopped first at the tiny station at Grove Ferry and later at Chislet Colliery Halt, a specially constructed station to serve the colliery. As late as the 1960s and 70s special buses ran from the Thanet resorts and from as far as Herne Bay on the north Kent coast to take miners to work at Chislet.

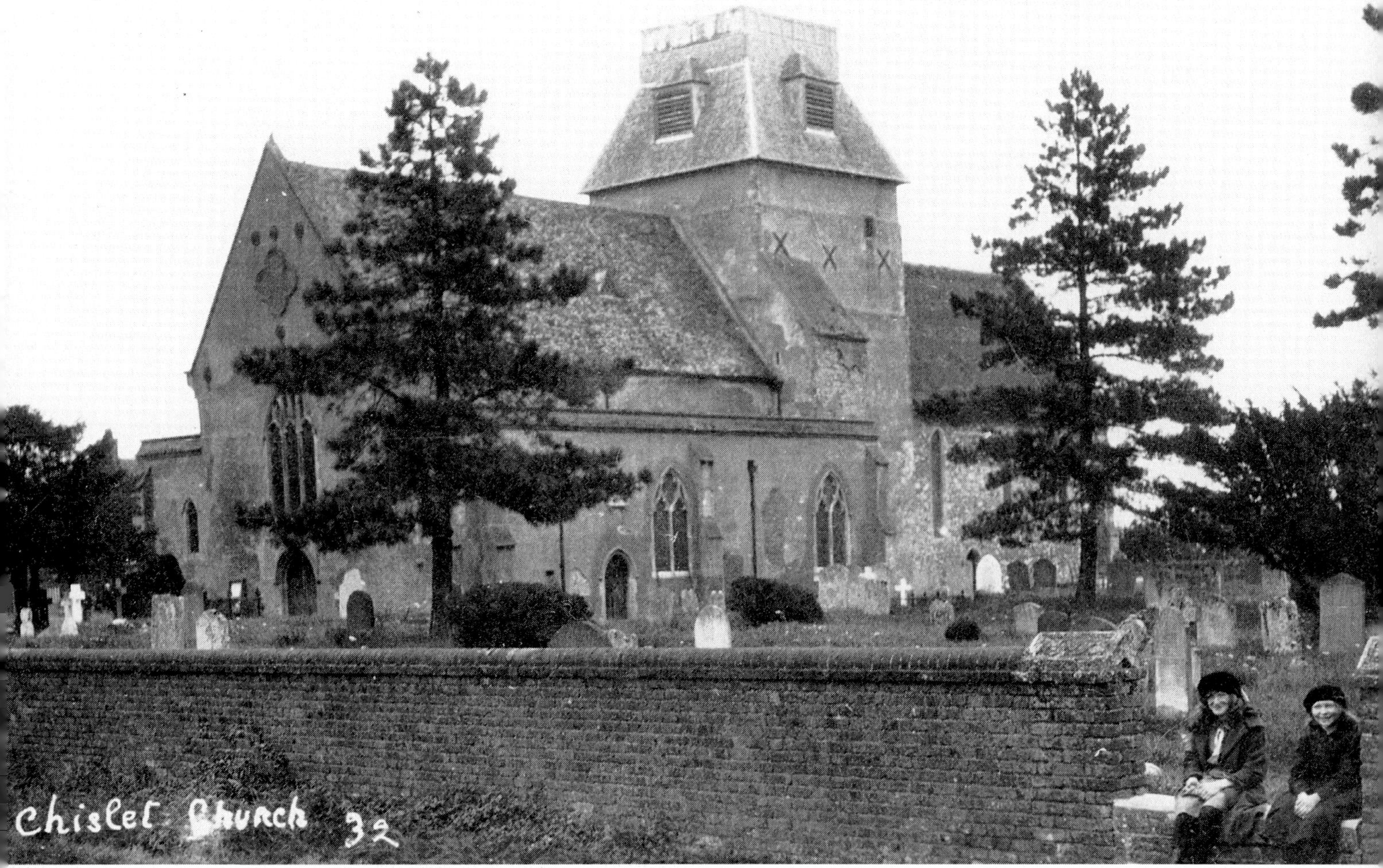
Chislet Church 32

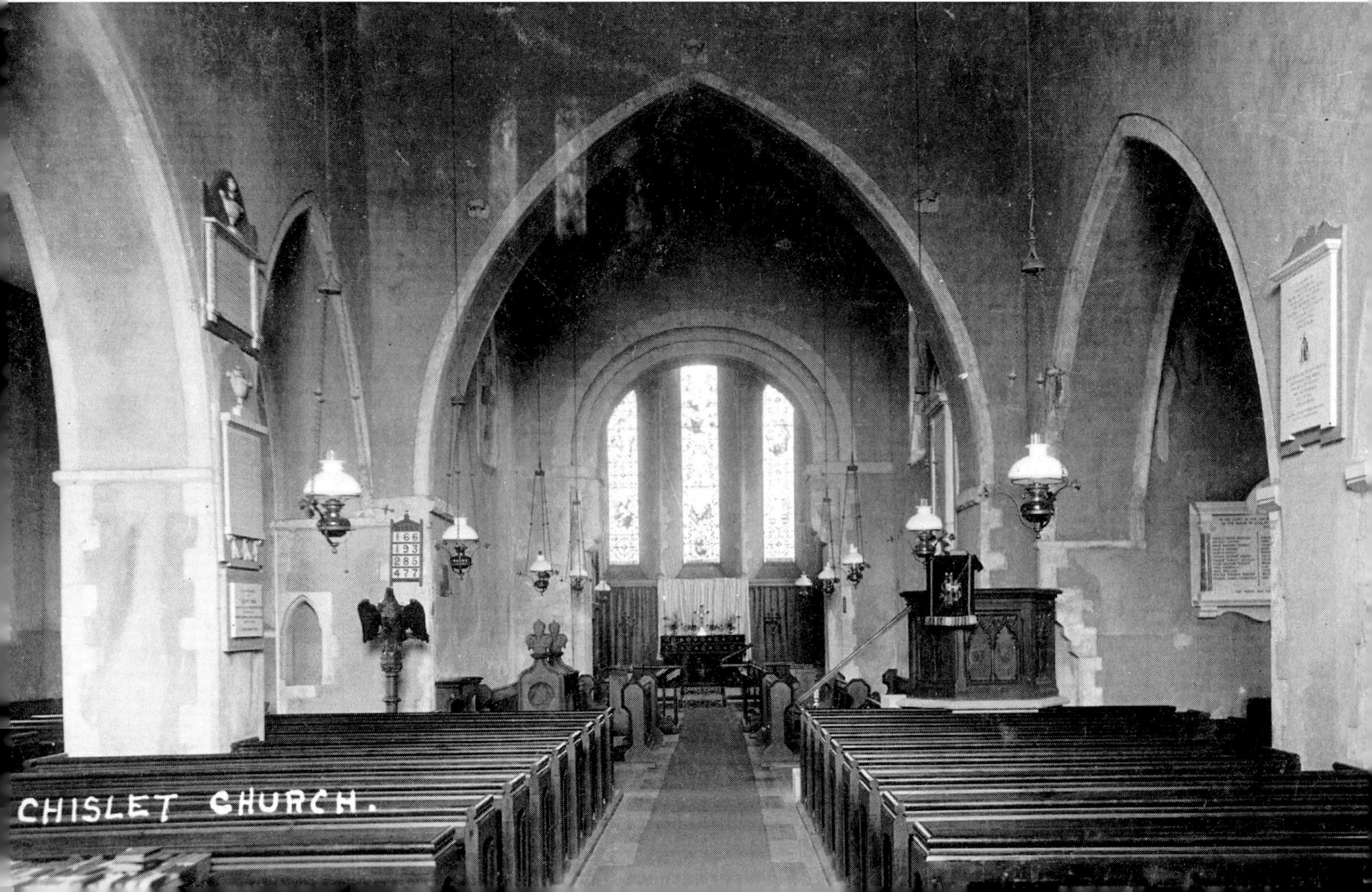
166
193
285
477
CHISLET CHURCH.

Inside new bath houses Chislet.

The actual village of Chislet is very ancient, dating back to at least 600 A.D., while in Roman times there was an established settlement in the area. In 1924 the Chislet Colliery Housing Society came into being, its remit to build a village of 300 houses close to the colliery. This estate was later named Hersden to distinguish it from the old Chislet Village. In 1924 Chislet Colliery opened its pit head baths. the first in Kent. During the Second World War Chislet was one of many collieries chosen to be training sites for the young boys and men who had been conscripted to work in the mines rather than to join the armed forces. Soon after the war began the government had realised that there was an acute shortage of miners and they knew a solution had to be found. The shortage of coal miners during the war was the result of the government of the day 'calling-up' miners to serve in the armed forces without stopping to think who would be left to mine the very important coal needed to power factories engaged in manufacturing equipment for the war. The young men who subsequently went to work in Britain's coal mines were known as 'Bevin Boys', after Ernest Bevin, the Minister of Labour and National Service. These young men were picked by ballot to work in Britain's pits rather than to enter one of the armed services. The scheme was started in December 1943 and involved a total of almost 48,000 young men picked at random, that number amounting to almost 10% of all conscripts. The first trainees to arrive at Chislet, 39 young men, were housed in a specially built hostel in the grounds of Wildwood, at Sturry, an already established facility that provided single miners with board and lodging. Most of the 'Bevin Boys' did a difficult job under even more difficult conditions. They were the butt of many jokes from the older miners, but their biggest problem was that they were not given a uniform or any form of identity documents that would make them known as miners as were soldiers, airmen or sailors. All they were issued with was a pair of boots and a helmet. As a result it was quite common for Bevin Boys to be stopped by over-zealous police as suspected deserters or insulted by members of the public who thought they were avoiding doing their duty by not being in uniform. The Bevin Boys received six weeks training, some in the classroom and some on site. They were also given PE classes, presumably to build up their muscles and get them as physically fit as possible.

The Bevin Boys did not have an easy time financially either. Although they were paid for their work, they were responsible for paying for their bed and board and the cost of travel from their homes to the area where they had been assigned to work. As they were, after all, conscripted workers, this was thought to be very unfair as members of the armed forces were afforded free board and lodging plus travel passes. When the war ended in 1945 not all the Bevin Boys returned to their previous occupations and were still kept working in the pits for several more years. It is only in the past year that they have finally been recognised by the government and awarded medals for working in the coal mines during the war. Many, of course did not gain the recognition they deserved because they had died in the intervening years.

Hersden Village, 1994.

March 1945 saw the arrival of the first pit ponies; they arrived by train and were unbroken. The miners fenced in an area of pasture and work was started on stables below ground to house them. The stables are said to have been very comfortable, built of wood, whitewashed and supplied with electricity. Each pony had his name on a board above his stable and all the ponies' names, for some reason, began with 'A'. The object of introducing ponies was supposed to be to lighten the amount of heavy work undertaken by the miners, but they had such an efficient system worked out already that the ponies were not really needed. The final pony to work at Chislet, Alex, was brought to the surface for the last time in 1952.

Hersden Village was quite isolated from the nearest town of Canterbury and as such a very cohesive community grew up there. Sports clubs to cater for all tastes were started and a silver band was formed under the direction of Jackie Robinson. The Chislet rugby team was formed by two enthusiasts, John Jones and Arthur Sutton, while a popular boxing club was started by a Scotsman, Dave Leitch. The local primary school was enlarged to cater for the influx of families and some shops were built to fulfil the needs of the new families. The ever popular Co-op provided a grocers and butchers shop.

Local shops at Hersden, *circa* 1950s.

Winding gear and first aid post, Chislet.

Social Club and Institute, Westbere.

Chislet Colliery

THE POST OFFICE. CHISLET.

Upstreet, Chislett.

Coal prep plant from main railway line.

In 1947 Chislet, along with all other collieries in Great Britain was nationalised and came under the control of the NCB. Coal had been placed under government control during both World Wars but ownership had stayed in private hands. In 1946 the Coal Industry Nationalisation Act was passed and from January 1st 1947, 'vesting day', all coal miners began working for the National Coal Board and a new era in the Kent coal field had begun.

At the time of nationalisation there were more than 700,000 employees working for the NCB. Miners saw nationalisation as public ownership, a small part of the socialist dream, and a great hope for better working conditions and pay. Unfortunately things did not work out that way. A Conservative politician's opinion of nationalisation was that it was '*a near impossible brief of combining public service with commercial efficiency*'.

Chislet Colliery's main customer was British Rail, but when the use of steam locomotives were withdrawn in favour of diesel or electric power in the mid to late 1960s the colliery was no longer thought to be commercially viable. In the three months leading up to closure the monthly totals of coal mined from Chislet were, 8,975 tons in March, 7,192 tons in April and 8,095 tons in May of 1969. On June 10th that year the NCB notified Lawrence Daly, the NUM general secretary, that the colliery was to close. British mined coal and Kent coal especially could never compete with coal from French and German collieries which were very heavily subsidised by their respective governments. Also opencast coal mined in the USA and Australia would always be available at much lower cost. By July of 1969 the colliery closed for the final time. Some of the miners transferred to the other Kent pits at Betteshanger, Snowdown and Tilmanstone but the closure of Chislet was just the thin end of the wedge for the future of coal mining in Kent.

It was the end of a long-established way of life and a financial blow to all the miners and their families living in an area that already suffered from high unemployment. However, since the closure of Chislet Colliery the village of Hersden has not curled up and died but has taken on new life, not least by the efforts of the people of Hersden who were not willing to see their home village suffer more loss. The Chislet Colliery Welfare Club is still going strong from its founding in the late 1950s and forms one of the central social points of the village. The Hersden Community Centre, once located in the old police house has purchased the former Methodist church with the help of money from the Single Regeneration Budget. The building was refurbished partly with National Lottery funding and the result of much hard work by the Hersden Neighbourhood Community Association and now offers many services to the villagers. There are breakfast and after-school clubs for children between the ages of four and eleven. An over 50s lunch club, art classes and coffee mornings are also popular and held regularly. In addition there is a purpose-fitted online centre for those wishing to learn IT skills.

CHISLET COLLIERY

In 2003 SEEDA (South East England Development Agency) purchased four hectares of derelict land giving a total site of twelve hectares. The aim was to allow the expansion of any Canterbury businesses that wished to take on larger premises. It was hoped this would also help in job creation in the area. These plans have now been realised and where the colliery once stood there is a small industrial site and the area of marsh land near to the River Stour where once the spoil tip spread its ugly tentacles is now a nature reserve.

One way of still making money from coal has been the manufacture of small decorative items such as model cats and model steam locomotives that are cut out of coal. These items can be found at gift shops around the county.

CHISLET BRANCH
NATIONAL UNION OF MINEWORKERS
(KENT AREA)

Dear *Ross*

As we are all about to go on our separate ways it seems fitting that we should say farewell to each other with a great deal of pride on this very sad occasion, the closing of Chislet Colliery.

Everyone, none more than ourselves, knows that the reason for the closure is not because of any lack of effort on the part of the Chislet men. The productivity levels achieved will stand against any and could continue to do so, but, this is not to be.

Let us all remember, that if the one condition laid down for us in our jeopardy meeting had been honoured by the National Coal Board we would not be witnessing the closure of this pit. Just one condition was laid down for us; we achieved it, only to find that others were attached as an afterthought in order to beat us and justify a closure.

Whatever others in high places plan and scheme for the future, one thing they will never be able to take away from us is the fact that we were Chislet men, and we met all of their demands with a dignity that they will never be able themselves to match. We will be their conscience in the future when they begin once again to double deal.

Farewell and good health to you and yours for the future.

Thank you all for affording us the honour of representing you. Finally, in the words of Longfellow:

> Let us then be up and doing,
> With a heart for any fate,
> Still achieving, still pursuing,
> Learn to labour and to wait.

G. Leader (Chairman).
C. H. Jones (Secretary).
R. Gowen (Treasurer).
J. Collins (Delegate).
P. Shevlin (Compensation Secretary).
C. D. Gould (Vice-Chairman).
W. Baker (Committee).
J. Hemmings ,,
R. Rees ,,
V. Jones ,,
I. Dixon ,,
G. Mills ,,

CHISLET COLLIERY

NATIONAL COAL BOARD (KENT)   B/S/T

                                          1/3, Waterloo Crescent,
                                          Dover,
                                          Kent.

To: Mr. I. Llewellyn,
    19, The Avenue,
    Hersden,
    nr. Canterbury.

Check No.: 77   27th June, 1969

Dear Mr. Llewellyn,

In view of the fact that Chislet Colliery is to close, your present work there will not be available after 26th July, 1969.

Accordingly, I hereby give you 4 weeks' notice of termination of your Contract of Employment at Chislet Colliery on 26th July, 1969.

However, because we wish to do everything possible to keep you in the Industry, I am offering you, on behalf of the National Coal Board, an alternative job as a   Mining Apprentice at Tilmanstone Colliery

starting 11th August, 1969.

You already know the locality of this colliery and I hope enough information is given about the job itself. Details about transport to the new colliery will be given to you in due course. If, however, you require further information, please come and see me (or one of my staff) who will be present at Chislet Colliery most days up to the Annual Holidays and can be contacted through the time office.

The conditions of service regarding hours of work, holidays, rest days, sick pay and injury benefit, pension scheme and the amount of notice required to terminate your Contract of Employment are standard throughout the Board, and remain as at present. For particulars of your terms of employment under these headings you may examine the Schedule of Terms of Employment for Mineworkers, which is available for inspection in File C.E.A. No. 1 at your colliery office.

If you accept or do not accept the offer, please use the attached proforma and indicate that the starting date is suitable.

I must ask you to let me have your acceptance within seven days, i.e. by 5th July, 1969.

                                          Yours sincerely,

                                          Idris Davies

                                          (Idris Davies)
                                          Industrial Relations Officer
                                          & Staff Manager

Overleaf: Jack Blackhurst on 1s unit.

BETTESHANGER COLLIERY

Located just off the A258 Deal to Sandwich Road Betteshanger was the largest colliery in the Kent coalfield. Coal was initially discovered in the Lydden Valley just north of Deal prior to the First World War at a depth of 1,476 feet. That boring, actually made at Northbourne, was to prove the forerunner of Betteshanger Colliery. The early exploration conducted in 1911 was initiated and paid for by Arthur Burr. Following the end of the war Dorman Long & Co. bought the mineral rights to a large portion of land to the west and north of Deal and following the purchase Dorman Long merged with Messer. S. Pearson & Sons to form Pearson & Dorman Long Ltd.

A railway line was built to serve the planned colliery and the first shaft was dug in 1924. There were plans to create further collieries at Wingham, Fleet, Woodnesborough, Stodmarsh and at Deal, but these plans never came to fruition. As with other pits in the Kent coalfield Betteshanger was at first beset by flooding and just as at other sites the cementation process that had been improved upon through the years was used to good effect and the sides of the shafts were sealed. The shafts at Betteshanger were 24 feet in diameter, the largest in Kent.

First coal, Betteshanger, 1924.

Progress was swift and the first seam of coal was reached in 1927. This was fortunate timing for both employers and miners. Following the General strike of 1926 there were many miners from traditional coal mining areas who had been blacklisted by the colliery owners and could no longer work in their home areas. These men came to the new collieries of Kent often walking hundreds of miles for lack of the train fare. Many, because of their employment status gave false names when they signed on at Betteshanger.

Aerial view of Betteshanger, 1930s.

At first the miners came without their families. Almost overnight Deal and its environs were overwhelmed by the arrival of some 1,500 men looking for lodgings in a small and somewhat genteel seaside resort. They were not met with open arms. As in the other towns and villages of east Kent the newcomers were perceived as rough and dirty, they spoke with accents the locals could not understand and they were not trusted. Signs went up in the windows of boarding houses, pubs and cafes, 'no miners'. When later the families of the miners arrived things just got worse. One miner walked through Deal pushing a pram containing his family's belongings and carrying a sign that read Home Wanted. There was no room for the children in local schools, medical facilities were barely adequate for the local population and the local traders took advantage of the miners' wives by selling them poor quality foods at inflated prices.

BETTESHANGER COLLIERY

Winding gear, workshops and drilling derrick, Betteshanger, 1924.

BETTESHANGER COLLIERY

Early view of Betteshanger with miners' houses bottom left.

Houses to accommodate deputies were constructed close to the colliery, but it was not until 1929 that land was obtained in the area known as Mill Hill on the outskirts of Deal, where miners' housing was built. Unlike the other accommodation provided for miners and their families at the other collieries in east Kent the estate at Mill Hill was not located near an existing village. There was a total of 950 homes plus sports facilities and a social club. At the time it was built the Mill Hill Estate was on the eastern edge of the main town of Deal, but now in the 21st century it has been surrounded by more houses and is seen and accepted as just another part of the town. Because of its closeness to Deal, Mill Hill was never seen as a pit village in the way that Aylesham, Eythorne and Chislet were. Following the construction of the houses at Mill Hill pit head baths were opened in 1934. In 1932 the Betteshanger Brass Band was formed and the Betteshanger Social and Welfare Sports Club opened in Cavell Square on the Mill Hill Estate.

Members of 1's team, Betteshanger. *Left to right*: **Dave Fraser, Jimmy Trice, Jack Blackhurst, Arnold Moyle, Dave Murphy.**

BETTESHANGER COLLIERY

Trainees at Betteshanger, 1956/57.

BETTESHANGER COLLIERY

London, 1960. This march was in support of 140 young miners who had been made redundant.

Because Betteshanger was worked by men who knew from previous experience how hard it was to fight intransigent bosses the colliery gained a reputation for being the most militant in Kent. In 1938 a strike was called over the mistreatment by a deputy towards a group of pit boys. Work was only resumed when agreement was reached to hold a public enquiry. A strike during the Second World War occurred regarding the allowances being made to work a particularly difficult seam. The result of the strike saw three union officials gaoled and 1,000 men being given the choice between paying a fine and performing hard labour. The government feared the strike would spread to other pits, a situation it could not tolerate in the midst of the war. All but nine of the miners refused to pay the fine. This left the authorities having to find almost a thousand prison places, an unrealistic task at any time. The miners knew they were needed at the coal face producing badly needed coal and not languishing in jail and in the end the government decided not to take any action. The imprisoned union men were also released.

Betteshanger produced a high quality coal that burned very hot, making it an ideal coking coal for use in the manufacture of steel. The spoil was transported from the mine by rail in the first instance but in 1974 an enclosed conveyor belt system was introduced. They conveyor carried the spoil beneath the Deal to Sandwich road and across farmland to an area known as Fowlmead; which was originally also agricultural land.

As was the case at other pits, the miners at Betteshanger saw Nationalisation as a good thing and their faith in the new management seemed to be borne out, at first. Money was put into improvements at the colliery and the miners felt they were winners in many ways. But it was not to be as they envisaged. They later realised that they were now 'under the thumb' of central government and that the NCB was just a branch of that government. The NUM discovered that not only was the NCB paying compensation to the pre-nationalisation owners of the mines, but that some industries were receiving cheap coal while the coal sold for domestic use was costing householders a great deal more. This did not go down well with householders who saw the high prices as the miners' way of getting pay increases, a view that was perpetuated by the media.

In the mid 1960s all workers, apart from face workers were being paid a common rate on a day wage basis while face workers still worked under a piece work system that was seen by the unions to be not only divisive and unsafe but also against the interests of both the workers and the industry. This arrangement resulted in a rash of local strikes, one of which at Betteshanger was held in the way of a 'stay down' by a group of miners. They had much local support including the Mayor of Deal, Alderman John Tapping, who went public in the local paper to record his backing of the miner's efforts. He also offered to act as mediator between the miners and the NCB. After six days the men were ordered to come up by the NUM. A standard rate was set by the NCB to be phased in over the next four years. Further disputes during the same decade resulted in a victory for miners when the NCB agreed to institute an eight hour day for all surface workers.

The 1960 miners' strike was however more about job security than pay. The NUM already knew that mining jobs were not infinite and they supported investment in the development of pits as a way of securing jobs for the future.

NCB Group Training Centre, Betteshanger.

The government's policy of selling coal below the commercial rates in the rest of Europe was estimated to have cost the coal industry two billion pounds in ten years. An industry that was in reality very profitable showed a loss purely because of government policy. The years of 1972 and 1974 saw much industrial upheaval. The strike of 1972 saw an improvement in miners' pay and conditions, but the increased rates of pay were being eroded by government policies on income. In November 1973, however, the NUM introduced an overtime ban and the government instituted a three day working week across the whole country to conserve coal stocks. In early February 1974, while plans for a strike were being made, Edward Heath the Prime Minister announced the dissolution of parliament to be followed by a general election on February 28th. The media was for the most part biased against the miner's actions and blamed them for the government crisis.

The Conservatives lost the election and the Labour Party came to power. The report on miners' pay was published recommending a rise in wages and other benefits such as extra pay for unsocial hours. These were implemented and the miners returned to work in March 1974. Investment was poured into coalfields and 200 million pounds was placed in a scheme to help sufferers of pneumoconiosis (a condition that causes fibrosis of the lungs brought about by working in an atmosphere contaminated with irritant dusts). Things, at least for the time being, were looking good, but even in 1981 a series of pits were marked for closure in the South Wales and Kent coalfields.

In 1984 the Conservatives under Margaret Thatcher came to power. An American industrialist, Ian McGregor, who had previously been in charge at British Leyland, was appointed as chairman of British Coal. He had a reputation for being tough and unbending in his actions. A plan of pit closures was put into effect and within days more than half the British coalfields were on strike, including those considered to be 'moderate'. The strike lasted for 16 months and was a bitter, violent and destructive period for British coal mines. One of the prime movers in the organisation of the actions taken by the government was Nicholas Ridley, Secretary of State for Transport at the time of the strike. Ridley had long seen that trade unions could be a threat to the execution of the Conservative government's policies. Following the debacle caused during Edward Heath's tenure in Downing Street, Ridley devised what would be known as the Ridley Plan. This covered preparations to be made well in advance of any miners' strike. Ridley had strong backing from Thatcher and by the time the strike began in March 1984 coal had been successfully stockpiled at power stations, and non-union workers had been put in place to ensure the working of oil-fired generation plants across the country. The Conservatives were prepared for a long fight.

Betteshanger Colliery

Some striking miners went underground during the strike to protest but all those men involved were charged with trespass and fired. At the end of the strike other miners at Betteshanger fought to get these men reinstated, but this was not achieved. During the sixteen months of the strike people from across the country raised funds to help the miners and their families. Support came in the shape of food and money from coal mining areas across Eastern Europe, countries which were then still a part of the Soviet Bloc. Several hundred support groups were set up, often led by miners' wives. These groups organised collections outside shops and ran communal kitchens with donated food. On one occasion the women from Kent joined with those from Yorkshire and organised a demonstration in Leicestershire to show support for the striking minority of miners in that area. Clothing, especially for children was donated and made available to miners' wives while concerts to raise money were given with all profits going to the miners. Some musical groups recorded songs about the strike and the proceeds of the sales went directly to miners' organisations. Rallies were held by students and other unions, but as the strike continued even the misrepresentation of the story by the media could not prevent ordinary people for seeing the strike for what it was, an all-out campaign not only to destroy the coal industry but to strongly curb all union activities throughout Great Britain. It was later revealed in her autobiography by Dame Stella Rimington the director of M15 (1992-1996) that M15 tapped the phones of NUM leaders during the strike. At the end of the 1984-85 strike the NCB became British Coal and as such, and with the blessing of the central government, proceeded to destroy the coal industry across the length and breadth of the country. In 1984 the British coal industry was the most productive in the world but even now when virtually all pits have been closed the coal reserves beneath the British Isles are still enormous. The strike finally came to an end on March 5th 1985, but because of the dispute arising out of the firing of the miners who staged a 'stay down' Betteshanger was actually the last pit to return to work.

Locomotive Battery Charging Station.

On the left is Neil Parker (Undermanager). Next to him on the right is Peter Yoxhall (Deputy Manager).

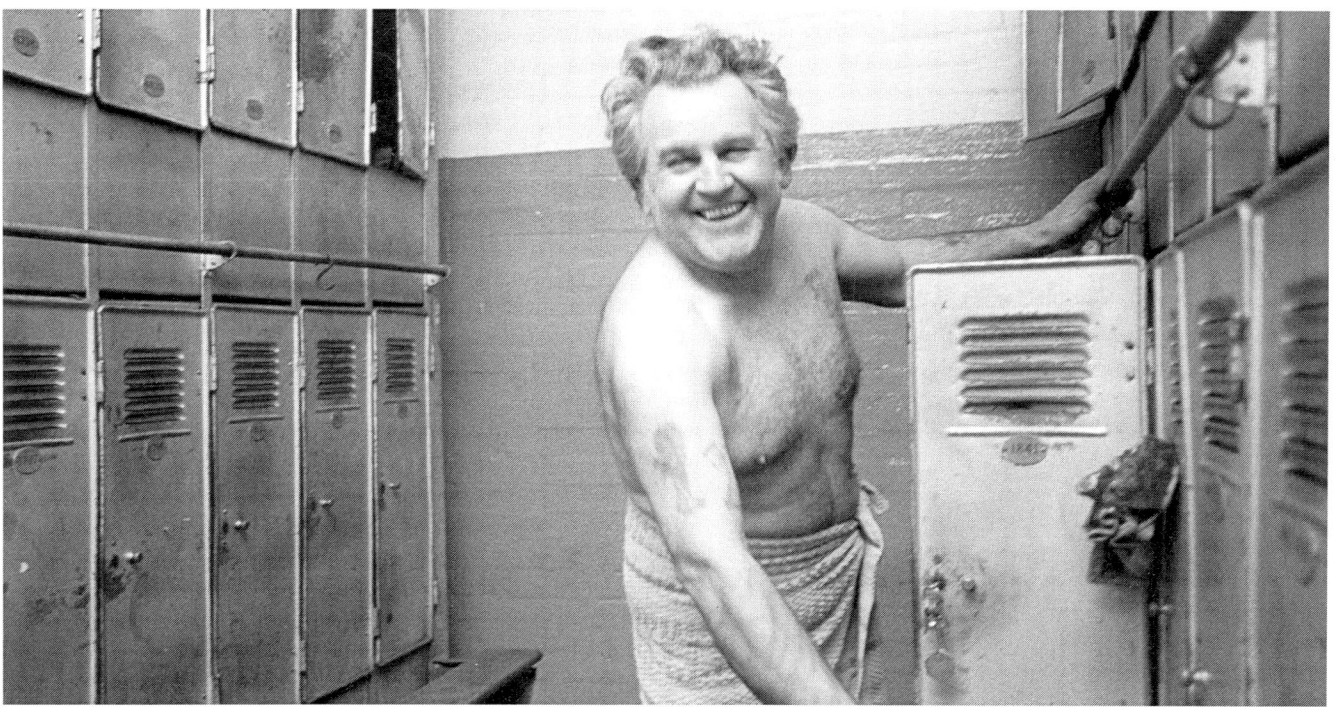

Tommy Mulanny getting ready to start his shift.

Throughout the strike Kent miners were among the most supportive of all miners across the country. At the beginning of the strike in November 1984 the percentage of miners on strike was 95.9%. In February 1985 the figure had only dropped to 95%. At the end of the strike this figure had fallen to 93%. Only the South Wales miners showed anything like a comparable solidarity. Betteshanger was the last colliery to close in Kent when it finally shut down production in 1989. It is a view strongly held by ex-Betteshanger miners that the pit was never developed to its full potential, this being mainly due to the economic errors and short-sightedness of central government, to say nothing of the underhand plans of many politicians. At one time proposals had been put forward to extend the workings eastward under the sea and to create further pits on the area of sandhills near to the Coach and Horses pub just off the Deal to Sandwich road. Coal measures had already been identified in that area many years before when borings had been made near Northbourne. Nothing ever came of these proposals and the rich coal seams are still there, an untapped energy resource that the country may well need to utilise one day. Unfortunately, as an ex-Betteshanger miner said recently, if it was ever considered pertinent to restart the mining industry all the equipment needed would have to be bought and imported to Britain because heavy industry in the UK has gone the same way as mining and the giant steel plants that made this country an industrial giant are now no more.

Following the pit closure not only did the men lose their livelihoods but their social clubs and sports facilities lost the financial support of the levy on miners' wages and British Coal. The social facilities previously enjoyed by the miners and their families were an important part of their lives; even with better transport links than in previous times some of the pit villages of Kent were still isolated. In the case of Mill Hill, however, the isolation was not that of distance but of class. Many residents across Deal still saw the miners, now actually ex-miners, as less than desirable neighbours. In the years since various new bodies have been set up to help in the regeneration of the area. The Coal Industry Social Welfare Organisation (CISWO), a charitable trust, was set up in 1995 and with their help the Betteshanger Social Welfare Scheme Sports Club was able to keep running. The Betteshanger Brass Band is still an important part of the community and has a full calendar of events throughout the year including appearances at the bandstand located on the sea front of nearby Walmer. The South East England Development Agency (SEEDA) is responsible for redeveloping the colliery site at Betteshanger, an area of 120 hectares. By 2005 the project had received £18.8 million pounds from English Partnerships. The colliery is now an industrial park and the site where the spoil tip once lay has been landscaped and was opened as a nature park, named Fowlmead, in May 2004 with a total of 130,000 shrubs and trees having been planted and protected areas set aside for birds, bats and lizards. A wildlife tunnel was built beneath the A258 to enable badgers and bats that are indigenous to area to safely move between the new business park and sites at Fowlmead. For cyclists there is a 3.5 kilometre circuit that has been described as the best in the country. There is also a visitor centre that incorporates a small café and a display of artefacts, photos and other information about the collieries. Special events are held to make visitors aware of the heritage of Kent's coalfields. SEEDA was also responsible for the setting up of the North Deal Community Partnership, a scheme of community renewal for an area where a lot of ex-miners live.

Overleaf, top row: **Colin Thorp, Jim Davies, Billy Grandville, Tony Morgan.**
Kneeling: **Steve Annel, Archbishop, Dave Rosser.**

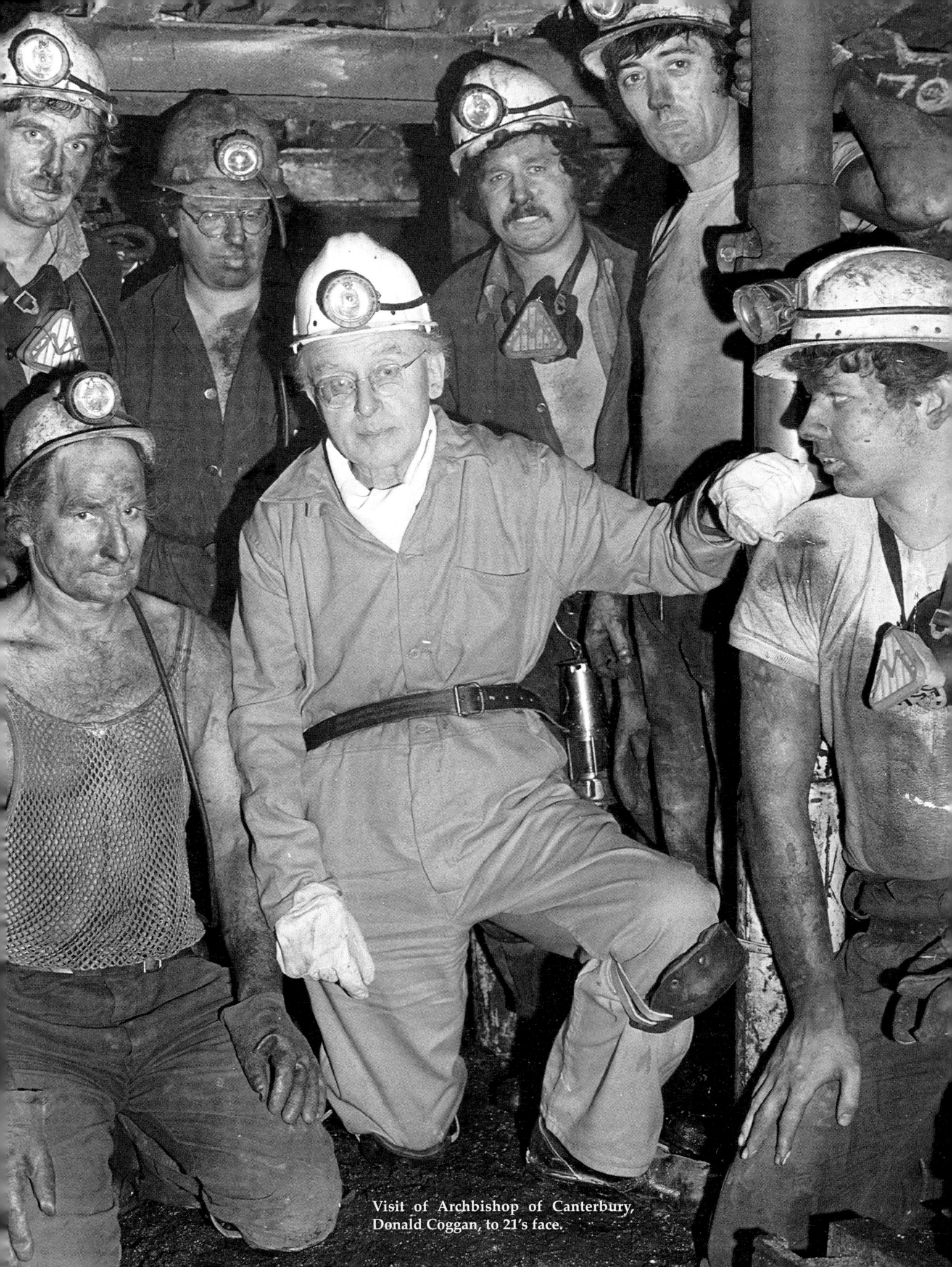

Visit of Archbishop of Canterbury, Donald Coggan, to 21's face.

BETTESHANGER COLLIERY

Jim McNicholas (Face Chargeman) and Dennis Crowley (Shearer Driver).

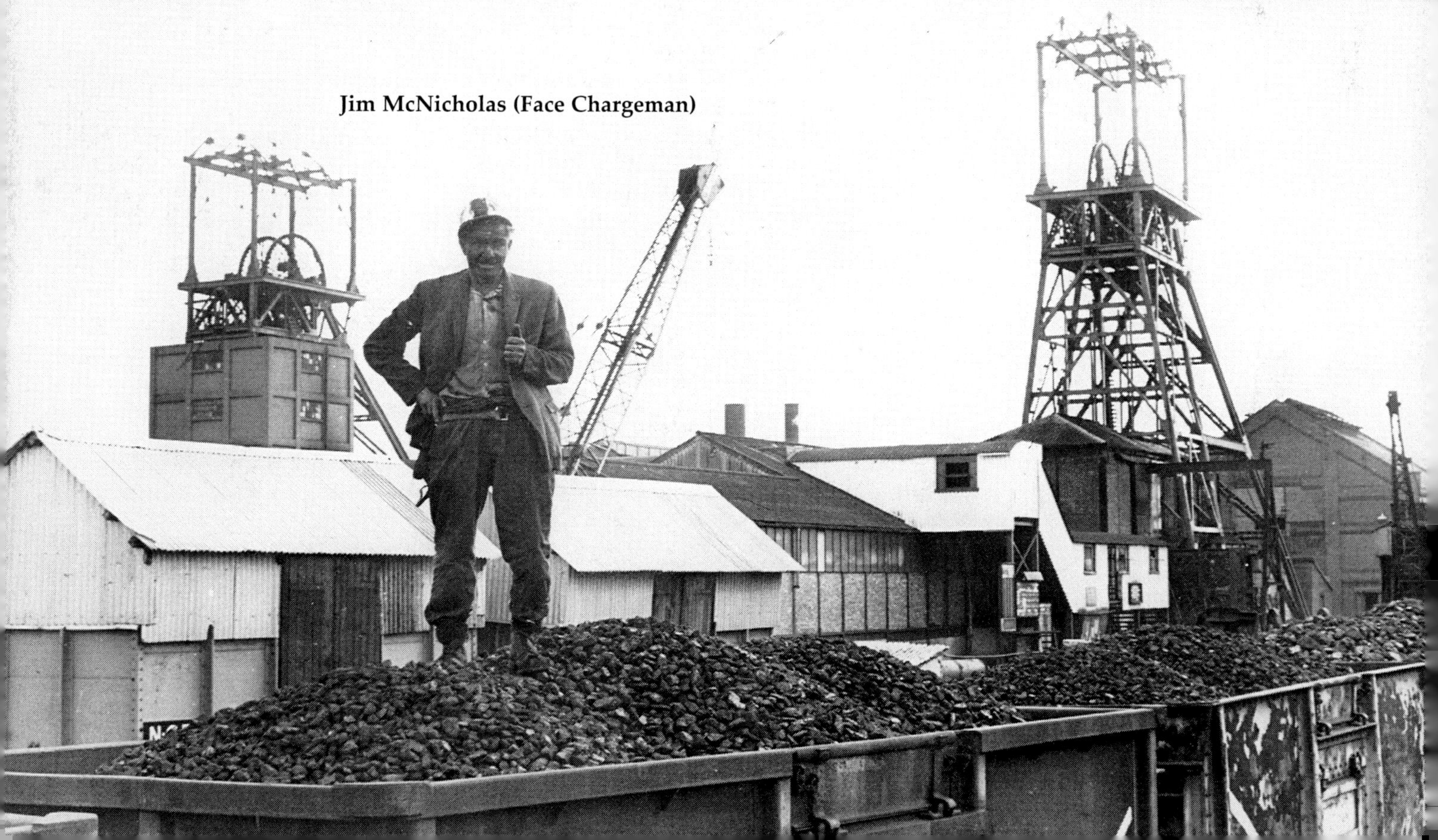

Jim McNicholas (Face Chargeman)

A Short Biography of Arthur Burr
Benefactor or Swindler?

No history of the Kent coalfield could be considered complete without a mention of Arthur Burr. With his cavalier methods of conducting business and an optimism that maybe even he believed in, Burr brought the sleepy east Kent countryside into the twentieth century with a bang. He was the archetypal Victorian speculator in a time when the population of these islands thought that nothing was beyond their achieving (God, after all was an Englishman). Burr took many risks and sailed close to the wind, but always comes across as a basically agreeable and friendly man. He was probably a very skilled con man or just could not bear to be seen to fail and appears to have been a very different person depending on who was giving their opinion of him.

Arthur Burr was born in Islington, London in 1850. He came from a comfortable middle class family and his father was a leather merchant. At the time of his birth both parents, William and Fanny were in their mid twenties. Arthur had two elder brothers and by the time he was a year old another brother had been born. The family kept three servants, including a nurse.

Burr first appears on records in Dover in 1905. By 1907 he was living with his wife and his son, Malcolm, in a very modest, terraced Victorian villa on Church Road in the Elms Vale area of the town and appears to have remained in that location throughout the rest of his life. That same year he occupied business premises at 58 Castle Street, just off the Market Place in Dover, which unfortunately were destroyed by enemy action towards the end of the Second World War.

By 1910 Burr had formed Kent Coal Concessions and was still conducting his business in Castle Street. Between 1905 and 1910 he had been the creator of no less than five collieries in east Kent and at one point, was either manager or director of 22 different companies. All the companies showed Kent Coal Concessions Ltd. to be the major shareholder. Many of the companies did not have boards of directors, just Arthur Burr in overall charge. In 1912 he became the (self-appointed) Director General of the East Kent Contract and Financial Company. By 1913 Burr's son, Malcolm, was involved in the business as Resident Engineer and was living at St. Margaret's Bay just to the east of Dover. He was responsible for the general supervision of Guilford, Tilmanstone and Snowdown collieries along with a Mr. Hollingworth, Consulting Engineer.

Burr's first venture into coal mining had been in 1896 when he formed the first Kent Coal Company. His goal was to build a colliery at Shakespeare Cliff. It was supposed to cost £50,000 and was predicted to produce 3,000 tons of coal a day by 1900. This never happened, the costs rose to more than £1million and coal was never obtained in any commercially viable quantity. Not to be deterred Burr convinced his investors of the viability of coal mining in Kent and went on to begin work on five more pits between 1904 and 1910. They were also not the commercial success Burr had predicted and none of the pits was able to produce sufficient quantities of coal until 1912. In spite of a lack of commercial success Burr told a meeting of the Dover Chamber of Commerce in October 1912 of the 'tremendous' profits that would attach to Kent coal being sold across the south east of England. When he attended a meeting of the East Kent Light Railway, Burr told them in no uncertain terms that if he could not get proper rates for the transportation of coal from the South East & Chatham Railways, the large main line rail company in Kent, he would go ahead and build his own railway from Kent to London to carry the coal.

Whatever Burr did in his business activities he had obviously impressed the mayor and councillors of Dover because, as early as October 1912, plans to honour him with a dinner were discussed at council meetings. He earned a reputation throughout Dover as being very much for the working man. During the winter of 1910-11 there was, apparently, a great deal of unemployment and distress among the workmen of Dover. When a friend phoned Arthur Burr and asked if he would subscribe to a fund that was being set up to help those in financial need Burr replied that there was no need for subscriptions to be taken up because he was prepared to take on practically all the unemployed, which he did. Burr was also responsible for forming a workmen's league. The mission of the league was to establish cordial relations between employees and employers, thus, hopefully obviating the occurrence of strikes in the future. The tradesmen in Dover concluded that the employment provided by Burr benefitted the town to the tune of £50,000 a year. At first the local paper, the *Dover Express*, strongly supported Burr's attempts to develop the Kent coalfield, but their views changed with time. In order to stay one step ahead of his creditors Burr indulged in a lot of what today would be called creative book-keeping. For almost twenty years he contrived to make the books balance. His method was to form new companies who would, on paper, appear to be subcontractors. One would own the mineral rights, another would sink shafts and yet another would be responsible for building winding houses, offices etc. on site. One of his companies could charge another for undertaking a certain job. It all became so complex that it is amazing that Burr was able to keep his businesses going for the best part of twenty years. The strain of all his dealings must have been unimaginable.

When word spread that the coal found in the Beresford seam at Tilmanstone was of excellent quality the share prices shot up, but at the same time rumours were rife that the coal was not of good quality at all. The origins of these opposing rumours is not known but, from what is known about Burr, it seems likely that the word of 'excellent' coal may have come from him.

Burr was constantly battling financial problems, with the shareholders on his back wanting to see returns on their investments and the press delighted in writing about his failures. However, to quash the derogatory rumours the following announcement was placed in the local paper in large eye-catching print.

£100 REWARD

For information about anyone known to be spreading rumours and false statements made with the intention of discrediting the companies with reference to Snowdown and Tilmanstone. Such rumours are without foundation, inspired by malice and stock exchange purposes. R. WREFORD Company Secretary Castle Hill House January 15th 1913 (Kent Coal Concessions & Allied Companies).

The whole episode has the ring of a publicity stunt and it is not recorded whether anyone ever tried to claim the reward.

To mark the successes at Snowdown Colliery a dinner was organised. Arthur Burr was the guest of honour and at the event he was presented with a Loving Cup in celebration of the raising of a workable colliery at Snowdown. The mayor predicted that the occasion of the dinner would be a 'Great day for the Kent Coal Company's shareholders'. The cup had been specially designed and its value was given as 250 guineas. The announcement continued with the invitation to all shareholders to be afforded the opportunity of visiting the collieries during the day. All particulars were to be embodied in a circular that was to be issued to all shareholders by Mr. Arthur Rowe, Honorary Secretary of Canterbury. At around the same time a meeting of Dover's town councillors discussed the probability of Mr Arthur Burr being awarded the Honorary Freedom of Dover. The Town Clerk read the resolution that was worded thus: *That the Town Council desired to place on record its sense of the eminent services rendered to the Borough by Mr. A. Burr in connection with the development of the East Kent coalfield, and the benefits that it directly conferred on Dover, by tendering him the Honorary Freedom of the Borough.* When the proposal was voted upon there was only one dissenting vote. The proposition that Arthur Burr be made a Freeman of Dover was carried in great detail by the *Dover Express*, indeed it ran to more than 26 double column inches. The mayor praised him as the greatest benefactor Dover had ever known and intimated very forcefully that without the assistance of Mr. Burr, Dover could not have survived the winters of 1910-11. It was agreed that the address should be embossed on vellum and contained within a casket. It is strange that in spite of his dubious business history Burr was given the Freedom of Dover for being 'one of the greatest benefactors Dover had ever known'. At a dinner given in Burr's honour the Sherlock Holmes' creator, Sir Arthur Conan Doyle was the guest speaker and he sang Burr's praises, going on to say that Dover was destined to be a 'Liverpool of the south' and one of the six biggest cities in Britain. One wonders if Conan Doyle held any shares in Burr's companies! The great occasion of Burr receiving the Honorary Freedom of Dover took place on February 4th 1913 and was probably the last high point of his very chequered career.

Finally his investors began to smell a rat and Burr was accused of being a conman. At the end of May 1914 a meeting was held in Folkestone at which a group of shareholders agreed to advance £30,000 to settle immediate bills on the condition that Arthur Burr and his son, Malcolm, both resign their positions along with the other members of the board. Burr was therefore forced out from all the posts he held and legal steps were started against him for fraud and misuse of funds, but this action did not appear to stop Burr's involvement with the coal companies. As late as 1918 Burr's business address given as 7, Castle Hill House reveals his continued involvement with coal exploitation to the tune of ownership of no less than nine companies all presided over by his secretary, Mr. J. Gurney.

In March of 1919 there was a meeting of the bankruptcy court to examine the financial affairs of Arthur Burr. The Official Receiver noted that Burr was unable to attend due to ill-health. At the time he was living at Kenilworth Court in Putney, London. After that time there is no more mention of Burr although there were other meetings of shareholders of his various companies. In the bankruptcy court the judge described him as a 'dangerous rogue' with more than £80,000 of judgements against him. When Arthur Burr died later that year he had numerous legal actions against him still pending.

After the death of his father Malcolm Burr appears to have had no further involvement in coalfield speculation in Britain. He carved out a successful career of his own as an author and well-known entomologist, travelling widely and recording his adventures, including a trip through Siberia. He died in 1954.

Share certificates of some of Arthur Burr's companies were recently auctioned. A 5/- share from 1907 signed by Burr as director of the East Kent Colliery Company was estimated to make £30 to £40. Share certificates from Burr's East Kent Contract and Financial Company Ltd. and his Foncage Syndicate Ltd. also went under the hammer.

RICHARD TILDEN SMITH
(The Consummate Capitalist)

Although Tilden Smith was the driving force behind what could be called the second phase of the development of the east Kent coalfield the comparison between him and his forerunner, Arthur Burr, could not have been greater. Burr was an adventurer and gambler and took risks with his own and other people's money. Such was his belief in the production of coal in east Kent that he lost sight of the main objective, to develop an industry that was built on a firm foundation with honestly obtained capital. He thought he was greater than the laws of the land and that just being enthusiastic was enough.

By contrast Tilden Smith was a professional business man, and brought with him to all of his business undertakings a wealth of experience gained in Australia. Born in New South Wales in 1865, the descendant of yeoman farmers from Sussex, his life followed a path of myth, coincidence and pure luck, the luck being the discovery of gold deposits. The outcome of such wealth saw the young Tilden Smith the owner of 5 million acres of land with 75,000 head of cattle. By the age of 21 he was in charge of the development of the Maitland Coalfield, the main source of coal in Australia.

In 1893 a bank crisis drastically reduced his finances and he arrived in England with little capital but the title deeds to a property that just happened to be located on the Kalgoorlie Goldfield. In England he established a creditable reputation for reconstructing businesses that were in danger of failing. The value of the companies he put back onto successful lines was estimated to be £200 million (at 1929 value). In 1908 Tilden Smith became a director of the Burma mines and also held the controlling interest in Chinese mines where a mining engineer by the name of Herbert Hoover was working, the same Herbert Hoover who became the 31st President of the USA. Tilden Smith eventually bought out Hoover's shares. The mines he now held were renowned for their rich sources of zinc, lead, copper and silver. At the start of the First World War, Tilden Smith took over the Swansea Vale Works and created a large, modern zinc smelting plant. In 1923 he sold the smelting works and his interests in the Burma mines. Before that, however, his interest in the Kent Coalfield had already been piqued and he became the owner of Tilmanstone in 1916.

Tilden Smith's main residence was in Brooke Street, Mayfair. He later moved from there and his wife occupied the house while he lived at a house close to London Bridge. When in Kent he lived at Flax Cottage within easy reach of his work. He and his wife had four daughters and one son. The son died during the First World War after being overcome by mustard gas in France. Although his personal life would have been the bread and butter of gossip columnists today he seemed unconcerned that his mistress was common knowledge. In fact she often accompanied him to Kent and was well-known by many workers at the colliery as his mistress. The opinion of the miners was that she was 'a real smasher!'

From personal opinions of the men who worked for him it would appear that Tilden Smith had a reputation for fairness and cared about the working conditions of those he employed, while at the same time being a hard working and thorough business man. He displayed vision and ideas, but in a practical way. When Richard Tilden Smith died in 1929 at the age of 64 he left so much undone. He had ambitious plans for Tilmanstone and indeed all of east Kent, but died suddenly on 18th December 1929 while at the House of Commons meeting politicians and promoting the coal industry in Kent. Following a lunch with MPs he collapsed and died as he was leaving the room. Two MPs who were doctors pronounced him dead at the scene. Because his death had occurred within a royal palace the coroner for royal palaces had to be notified. A post mortem found that Tilden Smith had died of heart disease and his own physician stated that he had been treating Tilden Smith for that condition for about twelve months. The physician had recommended that Tilden Smith should be 'very careful'.

He was survived by the four daughters and his estranged wife. The funeral was held on December 20th, leaving from his London residence, Adelaide House for the cremation at Golders Green Crematorium. The manager of Tilmanstone, Mr. Whittaker, the accountant Mr. Jenkins, and two men representing the miners attended the funeral.

His legacy to east Kent was an ongoing one, with recent extensions to Tilmanstone Colliery and the almost completed aerial ropeway. On the day of his death the ropes were connected along the seven and a half mile length of the ropeway from Tilmanstone to Dover Docks. In the village of Elvington he had overseen the construction of a further 100 houses. Had he not died so precipitously it is very possible that his plans would have seen the industrialisation of Kent and the growth of the port of Dover into one of the largest and busiest ports in Europe.